“Scientist of Islamic Era” is a book series encompassing eight volumes. The present book is volume 1 titled “Natural Scientists” that covers mathematics, astronomy, cryptoanalysis, chemistry, cartography, physics, and engineering based on these disciplines such as mechanics, automation, and robotics. The period of coverage is part 1 of Islamic Era, from AD 610 to 1400.

Many natural scientists in this book are mathematicians, astronomers, physicians and chemists at the same time, and also excelled in jurisprudence, hadith, philology, and poetry. They commanded exceptional breadth in their learning and deepest insights in their specializations; and, thus, greatly strengthened the foundations and expanded the frontiers of all fields of knowledge.

It is our objective that this first book in the series will inform the Muslims about the wealth of their scientific heritage, and the next generations will feel inspired to surpass the excellence of their ancestors to enrich their heritage further, and be, like their ancestors, the flag bearers of world civilization. It is also for the academic community to learn the truth about how science grew by leaps and bounds during Islamic era. And it is to quench the thirst of the youth, especially in Europe and Americas, to discover the truth about Muslim contributions to the world science, technology, and civilization; such quest has been greatly stinted by the historical biases and religious prejudices.

Muslims are now excelling in science and technology research with superb agility; our books in the series on “Scientists of Islamic Era” are expected to add impetus to this Renaissance in the Muslim world.

Natural Scientists

Volume 1 of the series on

Scientists of Islamic Era

Natural Scientists

Volume 1 of the series on

Scientists of Islamic Era

Abdur Rahim Choudhary

Muslim Voice

MV Publishers

Published by MV Publishers, a subsidiary of Muslim Voice,
12719 Hillmeade Station Dr, Bowie, MD 20720, USA.
MVPublishers@muslimvoice.org

ISBN 978-1-956601-04-6

First edition 2022
United States of America

Choudhary, Abdur Rahim, 1944–
Series on Scientists of Islamic Era, Volume 1, Natural Scientists / Muslim Voice

ISBN 978-1-956601-04-6

To the Muslim Ummah

Content

Preface to the series on Scientists of Islamic Era

For a period of more than a millennium, Muslim Scientists have done foundational research in all scientific disciplines, and also greatly expanded the frontiers of science. However, our people often do not have a clear idea about our scientific heritage. We decided to write a series of books on "*Scientists of Islamic Era*" that would be readily available to our generation and the coming generations, and provide motivation for excellence in the world civilizational dialogue, as well as to know our religious inspiration for scientific research and progression.

The young generations, especially those in Europe and Americas, have now opened their hearts and minds with a renewed desire for the truth about Islam and Muslims, being less influenced by historical biases and religious prejudices. The eight books in the series on *Scientists of Islamic Era* seek to serve their youthful thirst for the truth.

Another reason for this series on "*Scientists of Islamic Era*" is to produce a consciousness among the present-day academicians and scientists about the foundational contributions that the Muslim scientists made to all scientific disciplines, as well as how they expanded the frontiers of these disciplines. This fact is evidenced in the books in this series. However, this fact is not widely known because the present-day literature does not reference these original sources. The chain of scholarly references ends in European Renaissance, with occasional references to Greek scientists, but bypassing the millennium worth of research by Muslim scientists, who established the foundational principles and greatly expanded the frontiers of science.

In addition, the work seeks to fill a void, as no such series of books currently exists.

Islamic Era constitutes the period from 610 AD, when the Prophet received his first revelation, to 1922 AD, when the Ottoman Caliphate ended and the Turkish Republic began. We have divided the period in two parts: part 1 from 610 to 1400, and part 2 from 1400 to1922. The era is divided at an epoch when much of the works by the Muslim Scientists had already been translated into European languages, had become widely available, and had begun to produce Renaissance in Europe.

Each of the two parts of Islamic Era is covered by the following four volumes, eight volumes in all.

1. Volume 1 is for Natural Sciences that include mathematics, astronomy, cryptoanalysis, chemistry, cartography, physics, and engineering based on these disciplines such as mechanics, automation, and robotics.
2. Volume 2 is for the Medical Sciences that include physicians, nurses, surgeons, herbalists, medical researchers, and medical writers.

3. Volume 3 is for the Social Sciences that include philosophers, historians, physical geographers, qadhis, and hadith narrators, as well as the conventional sociology, political science, management sciences, economics, business, trade, anthropology, and linguists.
4. Volume 4 is for the Religious Sciences that include analogists, mohaddasin (historical fact checkers), jurisprudents, mofassarin (Quranic exegetists), and spiritualists (sciences of the tariqas).

We present this series of books to the readers to share with them the wealth of scientific excellence that these scientists contributed to the world civilization; to bring awareness to the Muslim readers about their role as the torch bearers of science and civilization; to serve the upwelling thirst that the young generation have for the truth about Islamic civilization; and to urge the academicians and researchers of the world, especially the Europeans and Americans, to learn and celebrate the Muslim giants of science upon whose shoulders they stand, and without whom the present-day scientific achievements could not have been possible.

Researchers like Professor Fuat Sezgin have devoted their lives to investigate contributions of the Muslim scientists. He has edited 1600 volumes. Such foundational work is invaluable for projects like ours. For example, there is a detailed Wikipedia article reporting his works on the subject, and is available under GNU free document license. Of course, we have performed extensive and critical editing and reorganization in order to serve our community well, to inspire them and our coming generations, and inform them of their role as torch bearers of excellence in world science, technology, and civilization. Acknowledgment is also due to Professor Abdur Rahim Choudhary and Ms. Yasmeen Sultana Choudhary whose total dedication made the work possible.

The Muslim scientists lived an integrated life with no conflict between the religion and the scientific passions; and a question never occurred that their scientific passion somehow needed to be separate from their religious inspirations. This is also obvious from the fact that most scientists were themselves experts in Islamic jurisprudence, hadith and Quran. In reality, their scientific work was also their religious worship because Islam showed them the necessity to do science, provided the motivation for it, and supported their scientific passion by equating it with religious worship.

No wonder they achieved scientific excellence with amazing grace.

The scientists are listed in chronological order, allowing an opportunity to correlate scientific tides and ebbs with political and religious ups and downs.

They could have been ordered according to the significance of their scientific contributions; that, however, is problematic because it is difficult, if not impossible, to assess the importance of research and compare across different scientific disciplines within natural sciences.

The order could have been sequenced according to how well the scientists are known today; that too is problematic because not all excellent scientists are well-known today, and, those who are, generally are made famous by the European commentators, who often did not know their works in original Arabic, and did not reflect the actual significance of their research. The well-known-ness is fairly arbitrary. For instance, Omar Khayyam is celebrated today for his Rubaiyat, which was something he did on the side, while his real works were in mathematics, a fact that is largely obscured.

This series of books should add to the impulse that is now thrusting the Muslims into the world of science and technology with increasing excellence in their achievements, signaling that their own Renaissance has now begun.

Muslim Voice
Bowie, MD, USA.
July 29th, 2022.

Preface to the First Edition of Volume 1: Natural Scientists

"Scientist of Islamic Era" is a book series encompassing eight volumes. The present book is volume 1 titled "Natural Scientists" that covers mathematics, astronomy, cryptoanalysis, chemistry, cartography, physics, and engineering based on these disciplines such as mechanics, automation, and robotics. The period of coverage is part 1 of Islamic Era, from AD 610 to 1400.

In this first edition, 68 natural scientists are included. Most of them are at least equivalent in their research excellence to the works of the modern-day scientists whose research is pioneering enough for a Nobel Prize. Many are above that stature because they were polymaths, commanding that level of excellence in multiple areas.

Many natural scientists in this book are mathematicians, astronomers, physicians and chemists at the same time, and they also excelled in jurisprudence, hadith, philology, and poetry. They commanded exceptional breadth in their learning and deepest insights in their specializations; and, thus, they greatly strengthened the foundations, and also expanded the frontiers, of all fields of knowledge.

It is our objective that this first book in the series will inform the Muslims about the wealth of their scientific heritage, and the next generations will feel inspired to surpass the excellence of their ancestors to enrich their heritage further, and be, like their ancestors, the flag bearers of world civilization. It is also intended for the academic community to learn the truth about how science grew by leaps and bounds during Islamic Era. And it is offered to quench the thirst of the youth, especially in Europe and Americas, to discover the truth about Muslim contributions to the world science, technology, and civilization; until now such quest has been greatly stinted by the historical biases and religious prejudices in Europe.

Muslims are now excelling in science and technology research with superb agility; our books in this series are expected to add impetus to this Renaissance in the Muslim world.

Abdur Rahim Choudhary, Ph.D.
Bowie, Maryland, USA
arc@muslimvoice.org
July 29th, 2022.

Natural Scientists

Natural Sciences community in Islamic Era was dominated by the Muslim scientists; the European scientists during this time were virtually nonexistent, owing to Europe being in the "Dark Age". When they started to emerge a little before the European Renaissance, they did so based on the research works of the Muslim scientists done for the prior seven centuries, which had already been translated into European languages, and had become broadly available.

These facts are obvious even if one examines not the entire scientific works by the Muslim scientists but only a subset of those that had been very visibly translated into European languages.

This book describes 68 natural scientists from part 1 (610-1400) of Islamic era (610-1922), covering the disciplines of mathematics, astronomy, cryptoanalysis, chemistry, cartography, physics, and engineering based on these disciplines such as mechanics, automation, and robotics. Each scientist is briefly described. First, the name of the scientist is disambiguated, and an attempt is made to correct the misrepresentations all too common in the European translations. Salient scientific contributions of each scientist are briefly highlighted, a difficult task because of the fact that most of these scientists were polymaths. For each scientist we have also provided a biographical summary to help picture the motivations and opportunities for them to do their research, in addition to their love and craving for knowledge.

The list of 68 natural scientists, that are covered in this edition of the book, is given in the table below, in chronological order. Each entry in the table includes the year of death and a one-line description, including the name of the scientist, the time period in parenthesis, and the area(s) of specialization within the natural sciences.

Table of 68 Natural Scientists covered in this edition of the book.

750	Harbi al-Himyari (8th century), alchemist
777	Ibrāhīm al-Fazārī (d. 777), mathematician and astronomer
806	Muḥammad ibn Ibrāhīm al-Fazārī (d. 796 or 806), Muslim philosopher, mathematician and astronomer
816	Jabir ibn Hayyan (died c. 806–816), alchemist and polymath, pioneer of organic chemistry; may also have been Persian
831	Al-Asma'i (739, Basra – 831, Basra), pioneer of zoology, botany and animal husbandry
835	Al-Khayyat (c. 770–c. 835), astrologer and a student of Mashallah
850	Khalid Ibn Abd Al-Malik (850) 9th century Astronomer
850	Al-Khwarizmi (780-850) Polymath, Mathematics, Astronomy, Geography,
869	Al-Jahiz (776–869), historian, biologist and author
873	Al-Kindi (c. 801–873), Arab philosopher, mathematician, astronomer, physician and geographer
880	Al-Farghani (d. 880), astronomer, known in Latin as Alfraganus
887	Abbas ibn Firnas (810-887) polymath: inventor, astronomer, physician, chemist, engineer, Andalusi musician, and Arabic-language poet.

902	Thābit ibn Qurra (826–902), mathematician, physician, astronomer, and translator
912	Ahmad ibn Yusuf (835, Baghdad – 912, Cairo), mathematician
929	Al-Battani (850, Harran – 929, Samarra), astronomer and mathematician
930	Abū Kāmil Shujā' ibn Aslam (850–930), mathematician
945	Al-Hamdani (893–945), geographer, historian and astronomer
950	Al-'Ijliyyah, (10th-century), female maker of astrolabes
950	Al-Ṣaidanānī (10th century), astronomer
950	Ibn Wahshiyya (10th century), Arab alchemist and agriculturalist
967	Al-Qabisi (d. 967), astrologer and mathematician
980	Al-Uqlidisi (920–980), wrote two works on arithmetic, may have anticipated the invention of decimals
985	Ibn al-A'lam (d. 985, Baghdad), astronomer and astrologer
1007	Maslama al-Majriti (950–1007), astronomer, chemist, mathematician, economist
1009	Ibn Yunus (c. 950–1009), mathematician and astronomer
1035	Ibn al-Saffar (d. 1035), astronomer
1035	Ibn al-Samh (979–1035), mathematician and astronomer
1037	Abī al-Rijāl (Haly Abenragel) (d. 1037), astrologer, best known for his Kitāb al-bāri' fi ahkām an-nujūm
1040	Ibn al-Haytham (965–1040), physicist and mathematician
1050	Ali ibn Khalaf (11th century), astronomer
1050	Ibn Khalaf al-Muradi (11th century) mechanical engineer and inventor
1050	Al-Biruni (973-1050) Polymath, Mathematics, Astronomy, Physics, Chemistry, Geodesy, Anthropology
1050	Yusuf al-Mu'taman ibn Hud (11th century), mathematician
1061	Ali ibn Ridwan (988, Giza – 1061, Baghdad), astronomer and geometer with Khalid Ben Abdulmelik

1070	Said al-Andalusi (1029–1070), astronomer, historian and philosopher
1079	Al-Jayyānī (989–1079), mathematician and author
1085	Ibn Bassal (b. 1085, Toledo), botanist and agronomist
1087	Al-Zarqali (1028–1087), mathematician, influential astronomer, and instrument maker, contributed to the famous Tables of Toledo
1118	Al-Tighnari (1073–1118), agronomist, botanist, biologist
1122	Al-Tughrai (c. 1061–1122), physician and alchemist
1131	Omar Khayyam (1048-1131) Polymath, Mathematician, Astronomer, Historian, Philosopher, Poet
1134	Abu al-Salt (c. 1068–1134), astronomer, physician and alchemist
1138	Ibn Bajja (Avempace) (1085, Zaragoza – 1138, Fez), philosopher, astronomer, physician
1037	Muhammad al-Baghdadi (1050-1141), mathematician
1150	Ibn al-'Awwam (12th century, b. Seville), agriculturist and botanist
1150	Jabir ibn Aflah (1100–1150), astronomer and mathematician who invented torquetum
1150	Al-Dimashqi, Abu al-Fadl (12th-century), writer and economist
1166	Al-Idrisi (1099–1166), geographer and cartographer
1195	Ibn al-Kammad (d. 1195), astronomer
1204	Nur ad-Din al-Bitruji (d. 1204), astronomer and philosopher; the Alpetragius crater on the Moon is named after him
1206	Ismail al-Jazari (1136–1206), scholar, inventor, mechanical engineer, artisan, artist
1213	Sharaf al-Dīn al-Ṭūsī. (1135-1213) Iranian mathematician and astronomer
1250	Fibonacci (1170-1250), Mathematician from Pisa
1250	Jordanus de Nemore (1250) Italian Mathematician
1266	Al-Urḍī (d. 1266), astronomer
1268	Ibn 'Adlan (1187, Mosul – 1268, Cairo), cryptographer and poet

1274	Nasir al-Din al-Tusi (1201-1274) Persian polymath, architect, philosopher, physician, scientist, and theologian
1295	Hasan al-Rammah (d. 1295), chemist and engineer
1296	Al-Ashraf Umar II (1242, Yemen – 1296, Yemen), astronomer and ruler of Yemen
1312	Zayn al-Din al-Amidi (d. 1312 AD), Islamic scholar and inventor
1315	Ibn al-Raqqam (1250–1315), astronomer, mathematician and physician
1321	Ibn al-Banna' al-Marrakushi (1256, Marrakesh – 1321), mathematician, astronomer, Islamic scholar, Sufi, and astrologer
1359	Ibn al-Durayhim (1312 – 1359/62), cryptologist
1375	Ibn al-Shatir (1304–1375), astronomer, mathematician, engineer and inventor, worked at the Umayyad Mosque in Damascus, Syria, developed an original astronomical model
1380	Al-Khalili (1320–1380), astronomer who compiled extensive tables for astronomical use
1405	Al-Damiri (1344, Cairo – 1405, Cairo), zoologist
1418	Al-Qalqashandi (1355/56–1418), writer and mathematician
1447	Ibn al-Majdi (1359–1447), mathematician and astronomer

We expect that this list will be expanded in subsequent editions, as further research is carried out.

A brief description for each scientist is provided, each in a separate subchapter. The 68 subchapters, that follow, are each dedicated to a single natural scientist. Some chapters are short, while others are detailed. Information on these topics is not abundant because the existing research is at best sporadic, and is mostly championed by individuals or small groups. There is a strong need for more detailed studies, on sustained and institutional bases, on an expansive scale.

The present series of eight volumes is offered in this context. They are intended for the educational and research institutes, at national and international levels, to provide encouragement for further work focused along these lines.

1. Harbi al-Himyari

Ḥarbī al-Ḥimyarī

(Arabic: حربي الحميري),

was a scholar from Yemen, who lived between the 7th and 8th century AD.

Scientific Contributions

He is mentioned several times in the writings of the Chemist Jābir ibn Ḥayyān (died c. 806–816) who is considered the father of Chemistry.

Harbi al-Himyari is one of Jabir's teachers: he is the mentor for teaching Koran and mathematics to Jābir ibn Hayyān.

Al-Himyari is the author of Kitāb ar-Rawḍ al-mi'ṭār.

One of Jabir's lost works is attributed to Harbi al-Himyari's contributions to Chemistry; indeed, there existed in Jabir's time a written work attributed to Harbi al-Himyari.

Al Himyari was practically unheard of in the West until 1975. Most of Western accounts of Al Himyari are therefore not trustworthy.

Biographical Summary

He was from Yemen. Not much is known about his life history. He died around 806-816.

2. Ibrahim al-Fazari

Ibrahim ibn Habib ibn Sulayman ibn Samura ibn Jundab al-Fazari (Arabic: إبراهيم بن حبيب بن سليمان بن سمرة بن جندب الفزاري).

(died 777 CE),

was an 8th-century mathematician and astronomer at the Abbasid court of the Caliph Al-Mansur.

Note: His son Muhammad ibn Ibrahim ibn Habib ibn Sulayman ibn Samra ibn Jundab al-Fazari (Arabic: إبراهيم بن حبيب بن سليمان بن سمرة بن جندب الفزاري) (died 796 or 806) was also a philosopher, mathematician, and astronomer.

Scientific Contributions

Ibrāhīm al-Fazārī composed various astronomical treatises:

"on the astrolabe", "on the armillary spheres", and "on the calendar".

He translated the book: Az-Zīj ʿalā Sinī al-ʿArab (Astronomical tables according to years of the Arabs). It was completed around 750AD in collaboration with his son and another scientist, Yaʿqūb ibn Ṭāriq.

At the end of the eighth century, while at the court of the Abbasid Caliphate, he mentioned Ghana's geography as "the land of gold', which was the first such description.

Biographical Summary

Ibrāhīm al-Fazārī was an 8th-century mathematician and astronomer at the Abbasid court of the Caliph Al-Mansur (r. 754–775). He died in 777 AD.

3. Muhammad ibn Ibrahim al-Fazari

Muhammad ibn Ibrahim ibn Habib ibn Sulayman ibn Samra ibn Jundab al-Fazari

(Arabic: إبراهيم بن حبيب بن سليمان بن سمرة بن جندب الفزاري),

(died 796 or 806),

was a philosopher, mathematician and astronomer.

Note: His father Ibrāhīm al-Fazārī, was also an astronomer and mathematician.

Scientific Contributions

Muḥammad ibn Ibrāhīm al-Fazārī built the first astrolabe; he was a pioneer and positively unrivaled in his mastery of the astral sciences.

Along with Ya'qūb ibn Ṭāriq and his father (Ibrāhīm al-Fazārī) he translated many scientific books, including the book, "Az-Zīj 'alā Sinī al-'Arab" (Astronomical tables according to years of the Arabs).

He also wrote a long poem on astronomy, Qaṣīda fī 'ilm (or hay'at) al-nujūm (Poem on the science [or configuration] of the stars).

Al-Fazārī wrote Kitāb al-Miqyās li-'l-zawāl (Book on the measurement of noon), Kitāb al-'Amal bi-'l-asṭurlāb wa-huwa dhāt al-ḥalaq (Book on the use of the armillary sphere), and Kitāb al-'Amal bi-'l-asṭurlāb al-musaṭṭaḥ (Book on the use of the astrolabe).

Biographical Summary

Muḥammad ibn Ibrāhīm al-Fazārī was a mathematician and astronomer at the Abbasid court of the Caliph Al-Mansur (r. 754–775). He died in 796 or 806 AD.

4. Jabir ibn Hayyan

Abū Mūsā Jābir ibn Ḥayyān

(Arabic: أبو موسى جابر بن حيّان),

(born in Iran c. 721, died in Kufah c. 815),

is author of an enormous number and variety of works. His writings covered a wide range of topics: cosmology, astronomy, astrology, medicine, pharmacology, zoology, botany, metaphysics, logic, and grammar.

Scientific Contributions

Jabir is the father of Chemistry. His works contain the foundational knowledge on the systematic classification of chemical substances. His work also forms the foundational knowledge of Chemical processes to derive inorganic compounds (for example ammonium chloride) from organic substances (such as plants, blood, and hair). He also established the sulfur-mercury theory of metals.

Jabir's research was influenced by a philosophical theory, ilm al-mīzān (the science of balance), which was aimed at reducing all phenomena (including material substances and their elements) to a system of measures and quantitative proportions.

Jābir's Seventy Books were translated into Latin as the Liber de septuaginta by Gerard of Cremona in the 12th century. A mutilated version of this work was known to the Latin pseudepigraphy practitioner who called himself Geber (transliterated from the Arabic Jābir), who wrote the Summa perfectionis magisterii (The Sum of Perfection or the Perfect Magistery), possibly the most famous alchemical book of the Middle Ages. It is probably composed in the late 13th century by a Franciscan monk known as Paul of Taranto.

Biographical Summary

Jabir lived in the 8th century and was a disciple of the Shi'ite Imam Jaʿfar al-Ṣādiq (died 765). His life and works are listed in the "Al-Fihrist" by Baghdadi bibliographer Ibn al-Nadīm (c. 932–995).

Some Western writers like to conjecture that Jabir Ibn Hayyan did not exist as a real person, basing it on some rumors.

5. Al-Asma'i

ʿAbd al-Malik ibn Qurayb al-Aṣmaʿi

(Arabic: أبو سعيد عبد الملك ابن قريب الأصمعي),

(739, Basra – 831, Basra),

was a pioneering biologist.

Scientific Contributions

ʿAbd al-Malik ibn Qurayb al-Aṣmaʿi pioneered studies in zoology, botany, animal husbandry and animal-human anatomical sciences.

He was also a great polymath scientist writing prolifically on philology, poetry, genealogy, and natural science.

He was an early philologist and one of three leading Arabic grammarians of the Basra school. He composed an epic on the life of Antarah ibn Shaddad.

His ambitious aim to catalogue the complete Arabic language in its purest form, led to a period he spent roaming with desert Bedouin tribes, observing and recording their speech patterns.

Asma'iyyat is Al-Aṣma'ī's magnum opus. It is a unique primary source of early Arabic poetry and was collected and republished in the modern era, by the German orientalist Wilhelm Ahlwardt. Al-Sayyid Muʿaẓẓam Ḥusain translated into English selected poems taken from both the Aṣma'īyyat and Mufaddaliyyat. It is important source of pre-Islamic Arabic poetry and is available online.

Most other existing collections were compiled by al-Aṣma'ī's students based on the principles he taught. Of al-Aṣma'ī's prose works listed in the Fihrist about half a dozen are extant. These include the Book of Distinction, the Book of the Wild Animals, the Book of the Horse, and the Book of the Sheep.

His Fuḥūlat al-Shu'arā is a pioneering work of Arabic literary criticism, And his Asma'iyyat is a compilation of an important poetry anthology.

Following is a list of his other writings.

- Disposition of Man or Humanity (كتاب خلق الانسان) - Kitab Khalaq al-Insan
- Categories (كتاب الاجناس)
- Al-Anwā' (كتاب الانواء) – "Influence of the stars on the weather"
- Marking with the Hamzah) (كتاب الهمز)
- Short and Long (كتاب المقصور والممدود)
- Distinction of Rare Animals (كتاب الفرق) - Kitab al-Farq
- Eternal Attributes [of God] (كتاب الصفات)
- Gates (كتاب الابواب) or Merit (كتاب الاثواب)
- Al-Maysir and al-Qidāḥ (كتاب الميسى والقداح)
- Disposition of the Horse (كتاب خلق الفرس)
- Horses (كتاب الخيل) - Kitāb al-Khail
- The Camel (كتاب الابل) - Kitāb al-Ibil
- Sheep (كتاب الشاء) - Kitāb al-Shā
- Tents and Houses (كتاب الاهبية والبيوت)

- Wild Beasts (كتاب الوحوش) - Kitab al-Wuhush
- Times (كتاب الاوقات)
- Fa'ala wa-Af'ala (كتاب فعل وافعل)
- Proverbs (كتاب الامثال)
- Antonyms (كتاب الاضداد)
- Pronunciations/Dialects (كتاب الالفاظ)
- Weapons (كتاب السلاح)
- Languages/Vernaculars (كتاب اللغات)
- Etymology (كتاب الاشتقاق)
- Rare Words (كتاب النوادر)
- Origins of Words (كتاب اصول الكلام)
- Change and Substitution (كتاب القلب والابدال)
- The Arabian Peninsula (كتاب جزيرة العرب)
- The Utterance/Pail) (كتاب الدلو)
- Migration (كتاب الرحل)
- The Meaning of Poetry (كتاب معانى الشعر)
- Infinitive/Verbal Noun (كتاب مصادر)
- The Six Poems (كتاب القسائد الست)
- Rajaz Poems (كتاب الاراجيز)
- Date Palm/Creed (كتاب النحلة)
- Plants and Trees (كتاب النبات والشجر)
- The Land Tax (كتاب الخراج)
- Synonyms (كتاب ما اتفق لفظه واختلف معناه)

- The Strange in the Ḥadīth (كتاب غريب الحديث نحو ماتتين ورقة رايتة بخط السكرى)
- The Saddle, Bridle, Halter and Horse Shoe (كتاب السرج و النجام والشوى و النعال *)
- The Strange in the Ḥadīth-Uncultured Words (كتاب غريب الحديث والكلام الوحشى)
- Rare Forms of the Arabians/Inflections/Declensions (كتاب نوادر الاعراب)
- Waters of the Arabs (كتاب مياة العرب)
- Genealogy (كتاب النسب)
- Vocal Sounds (كتاب الاصوات)
- Masculine and Feminine (كتاب المذكر والمؤنث)
- The Seasons كتاب المواسم

Al-Aṣma'ī was among a group of scholars who edited and recited the Pre-lslāmic and Islāmic poets of the Arab tribes up to the era of the Banū al-'Abbās.

He memorized thousands of verses of rajaz poetry and edited a substantial portion of the canon of Arab poets, but produced little poetry of his own.

Following is a list of poets that he edited:

- Al-Nābighah al-Dhubyānī (whom he also abridged)
- Al-Ḥuṭay'ah
- Al-Nābighah al-Ja'dī
- Labīd ibn Rabī'ah al-'Āmirī

- Tamīm ibn Ubayy ibn Muqbil
- Durayd ibn al-Ṣimmah
- Muhalhil ibn Rabī'ah
- Al-A'shā al-Kabīr, Maymūn ibn Qays, Abū Baṣīr
- A'shā Bāhilah 'Amir ibn al-Ḥārith
- Mutammim ibn Nuwayrah
- Bishr ibn Abī Khāzim
- Al-Zibraqān ibn Badr al-Tamīmī
- Al-Mutalammis Jarīr ibn 'Abd al-Masīḥ
- Ḥumayd ibn Thawr al-Rājiz
- Ḥumayd al-Arqaṭ
- Suhaym ibn Wathīl al-Riyāḥī
- Urwah ibn al-Ward
- 'Amr ibn Sha's
- Al-Namir ibn Tawlab
- Ubayd Allāh ibn Qays al-Ruqayyāt
- Muḍarras ibn Rib'ī
- Abū Ḥayyah al-Numayrī
- Al-Kumayt ibn Ma'rūf
- Al-'Ajjāj al-Rājaz, Abū Shāthā' 'Abd Allāh ibn Ru'bah. For his son, see Ru'bah.
- Ru'bah ibn al-'Ajjāj, called Abū Muḥammad Ru'bah ibn 'Abd Allāh, was a contemporary of al-Aṣma'ī whose poetry al-Aṣma'ī recited.

- Jarīr ibn 'Aṭīyah al-Aṣma'ī was among group of editors who included Abū 'Amr [al-Shaybānī], and Ibn al-Sikkīt.

Al-Aṣma'ī wrote some 60 works mainly on the animals, plants, customs, and grammatical forms. This reflects his interest and pioneering work in zoology, botany, animal husbandry, and human-animal anatomy; and also, philology.

Biographical Summary

Ibn Isḥaq al- Nadīm is a great bibliographer and biographer of Baghdad who compiled the encyclopedia Kitāb al-Fihrist (The Book Catalogue). In his 10th biography of al-Aṣma'ī he applies the science of Hadith to biography using the "isnad" narrative or 'chain-of-transmission' tradition. Al-Nadīm reports Abū 'Abd Allāh ibn Muqlah's written report of Tha'lab, giving Al-Aṣma'ī's full name as 'Abd al-Malik ibn Qurayb ibn 'Abd al-Malik ibn 'Ali ibn Aṣma'ī ibn Muẓahhir ibn 'Amr ibn 'Abd Allah al-Bāhilī.'

His father was Qurayb Abū Bakr from 'Āṣim and his son was Sa'īd. He belonged to the family of the celebrated poet Abū 'Uyaynah al-Muhallabī. Al-Aṣma'ī was descended from Adnān and the tribe of Bahila.

He was a student of Abū 'Amr ibn al-'Alā', the founder of the Basra school.

The governor of Basra brought him to the notice of the caliph, Harun al-Rashid, who made him tutor to his sons, Al-Amin and Al-Ma'mun. Al-Rashid once held an all-night discussion with al-Asma'i on pre-Islamic and early Arabic poetry. Al-Aṣma'ī was popular with the influential Barmakid viziers and acquired wealth as a property owner in Basra. Some of his protégés attained high rank as literary men. Among his students was the noted musician Ishaq al-Mawsili.

Remarks

Al-Aṣmaʿi is a polymath who presents a real challenge for someone who tries to classify him as a natural scientist, philologist, grammarian, teacher, or diplomat. We have classified him as a zoologist, notwithstanding his equally excellent credentials in other fields. We have nevertheless described his excellence in other areas as well, because he offers a especially difficult challenge by way of trying to place him in a bin.

As a matter of policy for this series on "Scientists of Islamic Era" we have decided not to multiple count a single scientist. As a result, we shall not include Al-Aṣmaʿi among the grammarians or philologists.

6. Al-Khayyat

Abu 'Ali al-Khayyat,

(c.770 - c.835),

was an astrologer.

Scientific Contributions

Abu 'Ali al-Khayyat was an astrologer. Two of his works are *Kitāb al-Mawālid*, "Book of Birth" and *Kitāb Sirr al-'Amal*, "Book of the secret of actions".

Biographical Summary

Not much is written about the life of Abu 'Ali al-Khayyat. There are anecdotal statements like he was a student of Masha'allah but no evidence of that is presented from the original sources.

7. Khalid Ibn Abd Al-malik

Khālid ibn ʿAbd al-Malik al-Marwarrūdhī

(Arabic: خالد بن عبدالملك المرو الروذي),

was a 9th-century astronomer from Baghdad.

Scientific Contributions

In 827 AD Khālid ibn ʿAbd al-Malik measured at 35 degrees north latitude, in the valley of the Tigris, the length of a meridian arc. This work he did in collaboration with ʿAlī ibn ʿĪsā al-Asṭurlābī. They measured Earth's circumference. They used a celestial horizontal coordinate system; and used the star altitudes to measure the geographical latitudes of the end points of the meridian arc.

The value they measured was 40,248 km. This was an accurate measurement of the size of the Earth (today's measurements yield a value of 40070 Km).

Biographical Summary

Khālid ibn ʿAbd al-Malik was a 9th century astronomer from Baghdad.

8. Muhammad ibn Musa al-Khwarizmi

Muḥammad ibn Mūsā al-Khwārizmī

(Persian: محمد بن موسی خوارزمی), or al-Khwarizmi,

(c. 780 – c. 850),

was a Persian polymath from Khwarazm, who produced vastly influential works in mathematics, astronomy, and geography. Around 820 CE, he was appointed as the astronomer and head of the library of the House of Wisdom in Baghdad.

Scientific Contributions

Al-Khwārizmī's contributed to mathematics, geography, astronomy, and cartography. He contributed pioneering works in algebra and trigonometry. He adopted systematic approach to solving linear and quadratic equations in his book on the subject, "The Compendious Book on Calculation by Completion and Balancing".

In his Calculations around 820, he used the decimal numeral system, also known as Arabic numerals, and spread the system of Arabic numerals throughout the Middle East and Europe. The quantum advancement in mathematics was translated into Latin as Algoritmi de numero Indorum, which introduced the Arabic numerals to Europe, who were archaistically using the primitive system of Roman numerals, which did not permit mathematics operations except very primitive ones using abacus.

Al-Khwārizmī wrote a major book "Kitab surat al-ard" ("The shape of the Earth". This presented the coordinates of places with improved values for the Mediterranean Sea, Asia, and Africa.

He also wrote on mechanical devices like the astrolabe and sundial. He assisted a project to determine the circumference of the Earth and in making a world map for al-Ma'mun, the caliph, which used a team of 70 geographers. When, in the 12th century, his works spread to Europe through Latin translations, it had a profound impact on the advancement of mathematics in Europe.

The Compendious Book on Calculation by Completion and Balancing (Arabic: الكتاب المختصر في حساب الجبر والمقابلة al Kitāb al-mukhtaṣar fī ḥisāb al-jabr wal-muqābala) is a mathematical book written approximately 820 CE. The book was written with the encouragement of Caliph al-Ma'mun as a popular work on calculation and is replete with examples and applications to a wide range of problems in trade, surveying and legal inheritance. The term "algebra" is derived from the name of one of the basic operations with equations like the operation of al-jabr (meaning "restoration", referring to adding a number to both sides of the equation to consolidate or cancel terms). The book was translated in Latin as Liber algebrae et almucabala by Robert of Chester (Segovia, 1145), and also by Gerard of Cremona. A unique Arabic copy is kept at Oxford and was translated in 1831 by F. Rosen. A Latin translation is kept in Cambridge.

The work provided an exhaustive account of solving polynomial equations up to the second degree, and discussed the fundamental methods of "reduction" and "balancing", referring to the transposition of terms to the other side of an equation, and the cancellation of like terms on both sides of the equation.

Al-Khwārizmī's method of solving linear and quadratic equations worked by first reducing the equation to one of six standard forms (where b and c are positive integers), as illustrated below.

- squares equal roots (ax2 = bx)
- squares equal number (ax2 = c)
- roots equal number (bx = c)
- squares and roots equal number (ax2 + bx = c)
- squares and number equal roots (ax2 + c = bx)
- roots and number equal squares (bx + c = ax2)

Equation is solved by dividing out the coefficient of the square and using the operations of al-jabr and al-muqābala.

It is important to understand just how significant this new idea was. It was a revolutionary move away from the Greek concept of mathematics which was essentially geometry. Algebra was a unifying theory which allowed rational numbers, irrational numbers, geometrical magnitudes, etc., to all be treated as "algebraic objects". It gave mathematics a whole new development path so much broader in concept to that which had existed before, and provided a vehicle for future development of the subject.

Algebra allowed mathematics to be applied to itself in a way which had not happened before.

The entire discipline of "Algebra" in mathematics derives from this monumental work. Similar is the situation regarding other works discussed below.

Al-Khwārizmī's another influential work was on the subject of arithmetic, which survived in Latin translations but is lost in the original Arabic. His writings include the text kitāb al-ḥisāb al-hindī ('Book of Indian computation'), and perhaps a more elementary text, kitab al-jam' wa'l-tafriq al-ḥisāb al-hindī ('Addition and subtraction in Indian arithmetic'). These texts described algorithms on decimal numbers that could be carried out on a dust board. Called takht in Arabic, a board covered with a thin layer of dust or sand was employed for calculations, on which figures could be written with a stylus and easily erased and replaced when necessary. Al-Khwarizmi's algorithms were used for centuries.

As part of 12th century wave of Arabic science flowing into Europe via translations of Arabic Science texts into Latin languages, these texts proved to be revolutionary in Europe, where they were still using the antiquated abacus.

Four Latin texts providing adaptions of Al-Khwarizmi's methods have survived, though none of them is believed to be a true translation:

- Dixit Algorizmi (published in 1857 under the title Algoritmi de Numero Indorum

- Liber Alchoarismi de Practica Arismetice
- Liber Ysagogarum Alchorismi
- Liber Pulveris

Dixit Algorizmi ('Thus spake Al-Khwarizmi') is the starting phrase of a manuscript in the University of Cambridge library, which is generally referred to by its 1857 title Algoritmi de Numero Indorum. It is attributed to the Adelard of Bath, who had also translated the astronomical tables in 1126. It is perhaps the closest to Al-Khwarizmi's own writing.

Al-Khwarizmi's work on arithmetic was responsible for introducing the Arabic numerals, to the Western world. The term "algorithm" is derived from the algorism, the technique of performing arithmetic with Arabic numerals developed by al-Khwārizmī. Both "algorithm" and "algorism" are derived from the Latinized forms of al-Khwārizmī's name, Algoritmi and Algorismi, respectively.

Al-Khwārizmī's Zīj al-Sindhind (Arabic: زيج السند هند, " is a work consisting of approximately 37 chapters on calendrical and astronomical calculations and 116 tables with calendrical, and astronomical data, as well as a table of sine values.

The original Arabic version (written c. 820) is lost, but a version by the Spanish astronomer Maslamah Ibn Ahmad al-Majriti (c. 1000) has survived in a Latin translation, presumably by Adelard of Bath (26 January 1126). The four surviving manuscripts of the Latin translation are kept at the Bibliothèque publique (Chartres), the Bibliothèque

Mazarine (Paris), the Biblioteca Nacional (Madrid) and the Bodleian Library (Oxford).

Al-Khwārizmī's treatise also contained tables for the trigonometric functions of sines and cosine. A related treatise on spherical trigonometry is also attributed to him. Al-Khwārizmī produced accurate sine and cosine tables, and the first table of tangents.

Al-Khwārizmī's third major work is his Kitāb Ṣūrat al-Arḍ (Arabic: كتاب صورة الأرض, "Book of the Description of the Earth"), also known as his Geography, which was finished in 833. It consists of a list of 2402 coordinates of cities and other geographical features following a general introduction.

There is only one surviving copy of Kitāb Ṣūrat al-Arḍ, which is kept at the Strasbourg University Library. A Latin translation is kept at the Biblioteca Nacional de España in Madrid. The book opens with the list of latitudes and longitudes, in order of "weather zones", that is to say in blocks of latitudes and, in each weather zone, by order of longitude. Neither the Arabic copy nor the Latin translation include the map of the world itself; however, it was reconstructed from the list of coordinates, the latitudes and longitudes of the coastal points in the manuscript. The points were transferred onto graph paper and connected with straight lines, obtaining an approximation of the coastline as it was on the original map. The same was done for the rivers and towns.

Ibn al-Nadim's Kitāb al-Fihrist, an index of books, mentions al-Khwārizmī's Kitāb al-Taʾrīkh (Arabic: كتاب التأريخ), a book of annals. No direct manuscript survives; however, a copy had reached Nusaybin by the 11th century, where its metropolitan bishop, Mar Elyas bar Shinaya, found it. Elias's chronicle quotes it from "the death of the Prophet" through to 169 AH, at which point Elias's text itself hits a lacuna.

Several Arabic manuscripts in Berlin, Istanbul, Tashkent, Cairo and Paris contain further material from al-Khwārizmī. The Istanbul manuscript contains a paper on sundials; the Fihrist credits al-Khwārizmī with Kitāb ar-Rukhāma(t) (Arabic: كتاب الرخامة). Other papers, such as one on the determination of the direction of Mecca, are on the spherical astronomy. Two other texts deserve special interest, on the morning width (Ma'rifat sa'at al-mashriq fī kull balad) and the determination of the azimuth from a height (Ma'rifat al-samt min qibal al-irtifā').

He also wrote two books on using and constructing astrolabes.

Biographical Summary

Few details of al Khwārizmī's life are known. Ibn al-Nadim gives his birthplace as Khwarazm, and he is generally thought to have come from this region. His name means 'the native of Khwarazm', a region that was part of Greater Iran, and is now part of Turkmenistan, and Uzbekistan.

Ibn al-Nadīm's Kitāb al-Fihrist includes a short biography on al-Khwārizmī together with a list of his books. Al-Khwārizmī accomplished most of his work between 813 and 833. After the Muslim conquest of Persia, Baghdad had become the center of scientific studies and trade, and many merchants and scientists from as far as China and India traveled there, as did al-Khwārizmī. He worked in the House of Wisdom established by the Abbasid Caliph al-Ma'mūn, where he studied the sciences and mathematics.

During the reign of al-Wathiq, he is said to have been involved in the first of two embassies to the Khazars.

Remarks

People talk about Algebra without necessarily digesting the meaning of this magnificent invention which took arithmetic way ahead. Along with numbers, discussed below, it also needs a similarly significant concept of negative numbers, which is usually just glossed over without bringing home its significance and impact.

Annotations

There is a good deal of confusion about the Indo-Arabic numbers: are they Indian or are they Arabic, who knew what of the decimal system, who discovered what?

One thing is clear, the Arabic Numerals allowed to get very significantly ahead of the Roman numerals, which the Europeans were using, till they learned the advanced system from Arab mathematicians.

The Romans put symbols together to form numbers; though they too were decimal in the sense that they introduced symbols for one, five, ten, hundred, thousand etc. They could form any number out of such symbols. There never was a need for a zero, because for every multiple of new decimal place they kept introducing a new symbol.

Indians did pretty much the same; for every multiple of ten they introduced a new name; ten, twenty, thirty, forty, fifty, sixty, seventy, eighty, ninety, hundred, thousand, ten thousand, lakh, ten lakh, karor, ten karor, arab, ten arab, kharab, ... The Indians knew how to count and they could write it down using these words. Still there was no need for a zero because they too introduced a new terminology at every multiple of ten.

So far both the Roman and the Indian systems were pretty much equivalent, the Roman system having a slight edge because they introduced names as well as symbols for multiples of ten, which facilitated writing them down compactly. However, neither had a way to do arithmetic manipulation except through mechanisms like the abacus.

Early in the seventh century, after the advent of Islam, the Indians thought of using a place holder through the introduction of a new symbol for zero. The symbols varied from an empty slot to a mere dot, etc.

The Muslims thought of doing the same.

The introduction of a zero by the Indians was very significant in principle, but in practice it did not help a lot. Though a place holder symbol for zero had been introduced, however its use to represent large numbers as a string of numerals was not grasped by the Indians.

This credit would go to the Muslims.

Europeans call them Arabic numerals because they learned them from translations of works originally written in Arabic. However, the authors of these works of Arabic texts were not always Arabs, though they were invariably Muslims. Al-Khwarizmi is one such example. Therefore, these numerals should more appropriately be called Islamic numerals.

The Muslims, therefore, pushed the arithmetic by a huge quantum leap; being able to write large numbers by writing a long string of digits (numerals). This opened the doors for arithmetic manipulation of numbers, through a process of string manipulations; calculating a new string from two given strings. Addition and subtraction became systematic processes, as described by Al-Khwarizmi in his book "kitab al-jam' wa'l-tafriq al-ḥisāb al-hindī" (Addition and subtraction in Indian arithmetic).

Here, the algorithm for addition and subtraction is entirely a Muslim invention, though they are totally honest in acknowledging the fact that the original idea of using a place holder zero came first from India.

Such advances in arithmetic then propagated back to India.

After the Muslims invented Algebra, they took arithmetic processes way ahead across many frontiers in a short period. They began to introduce numbers through powers of ten just as they had become experts in using powers of symbol x. The Arabs used the power manipulation processes in Algebra to solve complicated non-linear equations, systems of simultaneous equations, computing the nth root, and solving equations through plotting them into geometric curves. A parallel between the powers of x and the powers of 10 quickly allowed to handle astronomical quantities and calculations with those quantities.

9. Al-Jahiz

Abū ʿUthman ʿAmr ibn Baḥr al-Kinānī al-Baṣrī

(Arabic: أبو عثمان عمرو بن بحر الكناني البصري), commonly known as al-Jāḥiẓ (جاحظ),

(died 869 AD),

was a pioneering zoologist.

Scientific Contributions

Abū ʿUthman ʿAmr ibn Baḥr al-Kinānī al-Baṣrī commonly known as al-Jāḥiẓ, was a prose writer and author of works of literature, Mu'tazili theology, zoology, and politico-religious polemics. Ibn al-Nadim lists nearly 140 titles attributed to Al-Jahiz, of which 75 are extant.

Among the best known of his works is Kitāb al-Ḥayawān (the book of animals), which is a pioneering work on Zoology. It is a seven-part compendium on an array of subjects about animals.

Al-Jahiz clearly discusses the struggle for existence and natural selection. He also discusses micro evolution. Europeans ignore Al-Jahiz and attribute the theory of evolution entirely to Darwin. The work of Al-Jahiz precedes Darvin by a thousand years!

The theory of evolution should genuinely be renamed AlJahis-Darvin Theory.

Those who insist on Greek sciences as the source of Muslim works should note that the Greek had no inkling for a theory of biological evolution.

His other well-known works include Kitāb al-Bayān wa-l-tabyīn (the book of eloquence and Tabyin), a wide-ranging work on human communication; and Kitāb al-Bukhalāʾ (the book of misers), a book on stinginess.

Biographical Summary

His name is Abū ʿUthman ʿAmr ibn Bahr ibn Maḥbūb.

He worked with Abū al-Qallamas ʿAmr ibn Qalʿ al-Kinānī and ʿAmr ibn Qalʿ al-Kinānī al-Fuqaymī.

The grandfather of al-Jāḥiẓ was Mahbub, and his nephew reported that al-Jāḥiẓ's grandfather was a black cameleer (a camel driver).

His maternal grandfather was ʿAmr ibn Qalʿ nicknamed Fazārah.

Al-Jahiz asserted in a book he wrote that he was a member of the Arabian tribe Banu Kinanah.

Al-Nadīm reports that al-Jāḥiẓ said he was about the same age as Abu Nuwās and older than al-Jammāz.

Al-Jāḥiẓ died A.D. 869, during the caliphate of al-Muʿtazz.

Not much is known about al-Jāḥiẓ's early life. His family was very poor and he was a self-made man. He sold fish along one of the canals in Basra to help his family. Financial difficulties, however, did not stop al-Jāḥiẓ from continuously seeking knowledge.

He used to gather with a group of other youths at Basra's main mosque, where they would discuss different subjects.

Annotations

Learning was greatly facilitated by the cultural and intellectual revolution under the Abbasid Caliphate. Books became readily available, and learning greatly accessible.

Al-Jāḥiẓ studied philology, lexicography and poetry from among the most learned scholars at the School of Basra, where he attended the lectures of Abū Ubaydah, Al-Aṣma'ī, Sa'īd ibn Aws al-Anṣārī. He studied ilm an-naḥw (علم النحو, i.e., syntax) with al-Akhfash Abī al-Ḥasan.

While still in Basra, al-Jāḥiẓ wrote an article about the institution of the Caliphate. This is said to have been the beginning of his career as a writer. It is said that his mother once offered him a tray full of notebooks and told him he would earn his living from writing. He went on to write two hundred books in his lifetime.

In 816 AD Al-Jāḥiẓ moved to Baghdad, then the capital of the Abbasid Caliphate. That is because the caliphs encouraged scientists and scholars and had just founded the library of the Bayt al-Ḥikmah.

Al-Jāḥiẓ was said to have admired the eloquent literary style of the director of the library, Sahl ibn Hārūn (d. 859/860) and quoted his works. Because of the caliphs' patronage and his eagerness to establish himself and reach a wider audience, al-Jāḥiẓ stayed in Baghdad.

Al-Nadīm narrates from Abū Hiffān and the grammarian al-Mubarrad that al-Jāḥiẓ was reputed for being one of the three great bibliophiles and scholars; the other two being al-Fatḥ ibn Khāqān and judge Ismā'īl ibn Isḥāq. Al-Nadim continues that whenever a book came into the hand of al-Jāḥiẓ he read through it, wherever he happened to be. He even used the shops of al-warrāqūn for study.

Al-Jāḥiẓ replaced Ibrāhīm ibn al-'Abbās al-Ṣūlī in the government secretariat of al-Ma'mūn but left after just three days. Later at Samarra he wrote a huge number of his books.

The caliph al-Ma'mun wanted al-Jāḥiẓ to teach his children, but then changed his mind when his children were frightened by al-Jāḥiẓ's boggle-eyes (جاحظ العينين). This is said to be the origin of his nickname, al-Jahiz.

Al-Jahiz enjoyed the patronage of al-Fath ibn Khaqan, the bibliophile boon companion of Caliph al-Mutawakkil, but after his murder in December 861 he left Samarra for his native Basra, where he lived on his estate with his "concubine, her maid, a manservant, and a donkey." He died there in late 868, according to one story, when a pile of books from his private library collapsed on him.

10. Al-Kindi

Abu Yūsuf Ya'qūb ibn 'Isḥāq aṣ-Ṣabbāḥ al-Kindī

(Arabic: أبو يوسف يعقوب بن إسحاق الصبّاح الكندي),

(c. 801–873 AD),

was a philosopher, polymath, mathematician, physician and music theorist. He is the father of cryptography.

Scientific Contributions

Al-Kindi is the father of Cryptography. His book "Manuscript on Deciphering Cryptographic Messages" gave rise to the birth of crypt-analysis.

It is the earliest known use of statistical inference.

He also introduced several methods of breaking ciphers using frequency analysis.

Using his mathematical and medical expertise, he was able to develop a scale that would allow doctors to quantify the potency of their medication. This was the beginning of the employment of *quantitative analysis in Medical Sciences*, a foundational capability for the field of pharmacology.

Al-Kindi's research in optics set a milestone in the field. In a work known in the West as De radiis stellarum, al-Kindi developed a theory "that everything in the world emits rays in every direction, which fill the whole world." This theory of the active power of rays had an influence on later scholars such as Ibn al-Haytham, Robert Grosseteste and Roger Bacon.

Al-Kindi's philosophical writings demonstrated compatibility between philosophy and Islamic theology. These include such deep topics as the nature of God, the soul and prophetic knowledge. *He is the father of the philosophy of theology.* These fundamental ideas were soon further developed by al-Farabi.

Ibn al-Nadim' Fehrist contains two hundred and sixty books, contributing heavily to geometry (thirty-two books), medicine and philosophy (twenty-two books each), logic (nine books), and physics (twelve books). Although most of his books have been lost over the centuries, a few have survived in the form of Latin translations. However, some have been rediscovered in Arabic manuscripts; twenty-four of his lost works were located in the mid-twentieth century in a Turkish library.

Biographical Summary

Al-Kindi was born in Kufa in a family of the Kinda tribe. His father Ishaq was the governor of Kufa, and al-Kindi received his preliminary education there. He later went to complete his studies in Baghdad, where he was patronized by the Abbasid caliphs. On account of his learning and aptitude for study, al-Ma'mun appointed him to the House of Wisdom, a recently established centre in Baghdad. He was also well known for his beautiful calligraphy, and at one point was employed as a calligrapher by al-Mutawakkil; al-Mu'tasim appointed him as a tutor to his son.

In 873, al-Kindi died in Baghdad during the reign of al-Mu'tamid.

Remarks

Al-Kindi is a polymath who is equally excellent in mathematics, medicine, physics, philosophy, logic, and theology. It is an impossible challenge to fit him in any one place. In today's terms, he would win a Nobel Prize in all of these areas, hands down. Yet we have classified him as a natural scientist; and because of our policy for the series on "Scientists of Islamic Era" we would not double count him by also including him in other areas of knowledge. We are fully aware that we are thereby doing an injustice to Al-Kindi, and also to other areas of human enquiry. We will continue to research as to how best to treat this circumstance which actually applies to very many of the Muslim scientists.

11. Al-Farghani

Abū al-ʿAbbās Aḥmad ibn Muḥammad ibn Kathīr al-Farghānī

(Arabic: أبو العبّاس أحمد بن محمد بن كثير الفرغاني),

(800/805–870),

was one of the most famous astronomers.

Scientific Contributions

Al-Farghānī was one of the most famous astronomers. He worked in the Abbasid court in Baghdad. He composed numerous influential works on astronomy and astronomical equipment. Their influence was global because they were widely distributed in Arabic and Latin. His research documented in the treatise Kitāb fī Jawāmiʿ ʿIlm al-Nujūm (كتاب في جوامع علم النجوم) (Elements of astronomy on the celestial motions) is foundational. It contains new research, such as corrections to calculations of the circumference of the Earth, the Earth's axial tilt, and the apsides of the Sun and the Moon, as well as validated experimental measurements.

Al-Farghani also wrote several documents about astronomical instruments. His most famous is his treatise on the astrolabe, which is the oldest surviving document that details the theoretical construction and use of the tool. Al-Farghani's treatise on the astrolabe provides the mathematical basis for the construction of the astrolabe, along with tables containing thousands of data points enabling the construction of astrolabes that function at varying lines of longitude. It provides the mathematical justification for the functionality of the astrolabe.

Farghani was part of a team of scientists under the patronage of the ʿAbbāsid caliph al-Ma'mūn in Baghdad to calculate the diameter of the Earth by the measurement of the meridian arc length. This was the first such calculation.

In the 15th century, Christopher Columbus used al-Farghani's estimate for the Earth's circumference as the basis for his voyages to America. However, Columbus mistook al-Farghani's 7091-foot Arabic mile to be a 4856-foot Roman mile.

The Kitāb al-Fihrist by Ibn al-Nadim suggests that al-Farghani was also responsible for writing a book about the use and function of sundials.

Biographical Summary

Al-Farghani was born sometime in the early 9th century, and his last name might suggest that he is likely from Quva city, Farghana, though he has been described as Arab or Persian.

After his research in Baghdad he moved to Cairo, where he composed a treatise on the astrolabe around 856. There, he also supervised the construction of the large Nilometer, called the New Nilometer, on the Rawda Island (in Old Cairo) at the behest of the ʿAbbāsid caliph al-Mutawakkil, which was completed in the year 861.

This instrument allowed the height of the Nile to be measured in the event of a flood. This was first such monitoring system.

12. Abbas ibn Firnas

Abu al-Qasim Abbas ibn Firnas ibn Wirdas al-Takurini (Arabic: أبو القاسم عباس بن فرناس بن ورداس التاكرني), also known as Abbas ibn Firnas (Arabic: عباس بن فرناس), Latinized Armen Firman, (c. 809/810 – 887 A.D.),

was a Berber Andalusian polymath, an inventor, astronomer, physician, chemist, engineer, Andalusi musician, and Arabic-language poet.

He also experimented with flight.

Scientific Contributions

Ibn Firnas made various contributions in the field of astronomy and engineering. He constructed an astronomical device which indicated the motion of the planets and stars in the Universe. In addition, ibn Firnas came up with a procedure to manufacture colorless glass and made magnifying lenses for reading, which were known as reading stones.

Abbas Ibn Firnas devised a means of manufacturing colorless glass, invented various glass planispheres, made corrective lenses ("reading stones"), devised a chain of things that could be used to simulate the motions of the planets and stars, and developed a process for cutting rock crystal, which removed the need to send them to Egypt for cutting. He introduced the Sindhind to al-Andalus, which had important influence on astronomy in Europe. He also designed the al-Maqata, a water clock.

Al-Firnas was the first to attempt a flight. Some seven centuries after the death of Firnas, the Algerian historian Ahmed Mohammed al-Maqqari (d. 1632) wrote a description of Firnas that included the following:

> … one is his trying to fly. He covered himself with feathers for the purpose, attached a couple of wings to his body, and, getting on an eminence, flung himself down into the air, when according to the testimony of several trustworthy writers who witnessed the performance, he flew a considerable distance, as if he had been a bird, but, in alighting again on the place whence he had started, his back was very much hurt, for not knowing that birds when they alight come down upon their tails, he forgot to provide himself with one.

Biographical Summary

Abbas ibn Firnas was born in 810 AD in Ronda in the Takurunna province from Berber parents and lived in Córdoba. He died in 887.

13. Thabit ibn Qurrah

Al-Ṣābiʾ Thābit ibn Qurrah al-Ḥarrānī

(Arabic: ثابت بن قره),

(826 or 836 – February 18, 901),

was a mathematician from Mesopotamia, and lived in Baghdad in the second half of the ninth century; he was also an accomplished physician and astronomer.

Scientific Contributions

Thābit ibn Qurrah made important discoveries in algebra, geometry, and astronomy. In astronomy, Thābit is considered one of the first reformers of the existing theory. Thābit was a founder of statics in the mechanical science and engineering.

Thābit researched the problems of the motion of the sun and moon, and the theory of sundials.

Thābit determined the length of the sidereal year as 365 days, 6 hours, 9 minutes and 12 seconds (an error of 2 seconds). Sidereal year is when looking at the Earth and measuring it against the background of fixed stars, and it has a constant value.

In mathematics, Thābit discovered an equation for determining amicable numbers. He also wrote on the theory of numbers, and extended their use to describe the ratios between geometrical quantities; and worked on Transversal theorem in geometry.

He calculated the solution to a chessboard problem involving an exponential series.

He computed the volume of the paraboloid.

Thābit proved the Pythagoras' theorem through dissection, and described a generalization of this theorem. In regards to the Pythagorean Theorem, Thābit used a method based on reduction and composition.

Thābit proved Euclid's fifth postulate. In regards to Euclid's postulates, Thābit believed that geometry should be based on motion and more generally, physics. With that in mind, his argument was that geometry was tied with the equality and differences of magnitudes of such things as lines and angles.

He also wrote commentary on Archimedes' Liber Assumpta.

In physics, Thābit rejected the Peripatetic and Aristotelian notions of a "natural place" for each element. He instead proposed a theory of motion in which both the upward and downward motions are caused by weight, and that the order of the universe is a result of two competing attractions (jadhb): one of these being "between the sublunar and celestial elements", and the other being "between all parts of each element separately".

This is a formulation of a qualitative theory of gravity (without the inverse square law).

Thābit sought to establish a relationship between forces of motion and the distance traveled by the mobile.

This would latter give rise to Newtonian Mechanics with its laws of motion.

In mechanics, he was also a founder of statics.

In addition, Thābit proved the law of the lever.

Thābit's books include Kitab fi 'l-qarastun and Kitab fi sifat alqazn.

Biographical Summary

Thābit was born in Harran in Upper Mesopotamia, which at the time was part of the Diyar Mudar subdivision of the al-Jazira region of the Abbasid Caliphate. Some writers think that Thābit was a member of the Sabians community.

Thābit came to Baghdad to work with Banū Mūsā. They occupied themselves with mathematics, astronomy, astrology, magic, mechanics, medicine, and philosophy. Later in his life, Thābit's patron was the Abbasid Caliph al-Mu'tadid for whom he became a court astronomer. Thābit died in Baghdad in 901.

His son, Sinan ibn Thabit and grandson, Ibrahim ibn Sinan also made contributions to medicine and science.

Thābit wrote 150 works on mathematics, astronomy, and medicine. However, there are less than a dozen works by him that have survived.

Remarks

Thābit discovered a qualitative theory of gravity in astronomy; without the inverse square law, which would have required the notions of calculus.

Thābit also discovered a notional idea of Newtonian mechanics and contemplated on the laws of motion. To have completed the work would have required notions of calculus. However, the statics part of Newtonian mechanics did not require calculus and Thābit actually did formulate a theory of statics.

14. Ahmad ibn Yusuf

Abu Ja'far Ahmad ibn Yusuf ibn Ibrahim ibn Tammam al-Siddiq Al-Baghdadi

(Arabic: أبو جعفر أحمد بن يوسف بن ابراهيم بن تمام الصديق البغدادي),

(835, Baghdad–912, Cairo),

was a mathematician.

Scientific Contributions

It is not clear about some of his research if the text came from Ahmad ibn Yusuf, his father, or whether they wrote together. It is clear, however, that he worked on a book about ratio and proportion. This was translated to Latin by Gherard of Cremona. This book influenced early European mathematicians such as Fibonacci.

He also wrote a book on the astrolabe. He invented methods to solve tax problems that were later presented in Fibonacci's Liber Abaci. He was also quoted by mathematicians such as Thomas Bradwardine, Jordanus de Nemore and Luca Pacioli.

Biographical Summary

Ahmad ibn Yusuf was born in Baghdad and moved with his father to Damascus in 839. He later moved to Cairo: since he was also known as al-Misri, this probably happened at an early age. He died in Cairo in 912 AD.

His father worked on Mathematics, Astronomy and Medicine. He produced astronomical tables and was a member of a group of scholars.

15. Al-Battani

Abū ʿAbd Allāh Muḥammad ibn Jābir ibn Sinān al-Raqqī al-Ḥarrānī aṣ-Ṣābiʾ al-Battānī

(Arabic: محمد بن جابر بن سنان البتاني),

(c. 858 – 929),

was an astronomer.

Scientific Contributions

Al-Battānī proved a number of trigonometric relations.

His *Kitāb az-Zīj* was frequently quoted by many astronomers, including Copernicus.

One of al-Battānī's best-known achievements in astronomy was the determination of the solar year as being 365 days, 5 hours, 46 minutes and 24 seconds, which is only 2 minutes, and is 22 seconds off.

Another of al-Battānī's accomplishments is that he showed how an annular solar eclipse occurs. He did this by observing that the radius between the Earth and the Sun changes throughout the year. This led him to arrive to the conclusion that when the Sun is farthest from the Earth, an annular solar eclipse occurs. He was the first to make this observation and inference.

The twelfth-century Egyptian encyclopedist al-Qifṭī, in his biographical history Ta'rīkh al-Ḥukamā', mentions al-Battānī's contribution to advances in astronomical observation and calculations.

Al-Battānī amended some of Ptolemy's results and compiled new tables of the Sun and Moon, long accepted as authoritative. Al-Battānī observed that the direction of the Sun's apogee, as recorded by Ptolemy, was changing.

Some of Al-Battānī's measurements were more accurate than ones taken by Copernicus many centuries later.

Al-Battānī's work was instrumental in the development of science and astronomy. Copernicus, in his book that initiated the Copernican Revolution, the De Revolutionibus Orbium Coelestium, quotes his name no fewer than 23 times, and also mentions him in the Commentariolus.

Tycho Brahe, Riccioli, Kepler, Galileo and others frequently cited him or his observations. He also influenced Jewish rabbis and philosophers such as Abraham ibn Ezra, Abraham bar Hiyya, and Maimonides.

Al-Battānī's data is still used in geophysics. The major lunar crater Albategnius is named in his honor.

In mathematics, al-Battānī proved a number of trigonometrical relationships:

$\text{Tan } A = \text{Sin } A/\text{Cos } A$

$\text{Sec } A = \text{SQRT } (1 + \text{Tan}^2 A)$

He solved the equation sin $x = a \cos x$ discovering the solution:

$\sin x = a/\mathrm{SQRT}(1 + a^2)$

He also showed other trigonometric formulae for right-angled triangles.

Al-Battānī used al-Marwazi's idea of tangents ("shadows") to develop equations for calculating tangents and cotangents, compiling tables of them. He also discovered the reciprocal functions of secant and cosecant, and produced the first table of cosecants, which he referred to as a "table of shadows", in reference to the shadow of a gnomon in a sundial.

Using these trigonometrical relationships, Al-Battānī created an equation for finding the qibla direction.

Some of Al-Battānī's books are given below.

- Kitāb az-Zīj (كتاب الزيج or زيج البتاني, "Book of Astronomical Tables"); translated in Latin by many including De scientia stellarum. Bologna: Vittorio Benacci, eredi. 1645.
- Kitāb az-Zīj aṣ-Ṣābi' (كتاب الزيج الصابئ)

Translation published by Carlo Alfonso Nallino (1899-1907) under the Latin title *Al-Battānī sive Albatenii opus astronomicum: ad fidem codicis Escurialensis Arabice editum*; a multi-volume scientific treatise on geography and astronomical chronology from an Arabic manuscript with Latin annotations. The manuscript is held at the Escorial library.

- Arbaʿu Maqālāt (أربع مقالات, "Four discourses")

- Ma'rifat Maṭāli'i l-Burūj (معرفة مطالع البروج, "Knowledge of the rising-places of the zodiacal signs")
- Kitāb fī Miqdār al-Ittiṣālāt (كتاب في مقدار الاتصالات); treatise on the four quarters of the sphere.

Biographical Summary

Little of al-Battānī's life is known other than that he was born in 858 AD in Harran near Urfa, in Upper Mesopotamia, (today in Turkey); and that his father was famous as a maker of scientific instruments: Jabir ibn Sinan al-Harrani was likely this famous instrument maker.

Although his ancestors were likely of the Sabian-sect, he was a Muslim, as his name was Muhammad, with his kunya being Abu Abd Allah. The Encyclopedia Britannica Eleventh Edition stated that he had noble origins as an Arab prince, but *traditional Arabic biographers make no mention of this.* Between 877 and 918/19, over a forty-year period, he lived in the ancient city of Raqqa, in north central Syria, recording his astronomical observations.

He died in 929 AD at Qasr al-Jiss, which is located near Samarra, Iraq.

16. Abu Kamil

Abū Kāmil Shujāʿ ibn Aslam ibn Muḥammad Ibn Shujāʿ
(Arabic: أبو كامل شجاع بن أسلم بن محمد بن شجاع),
(c. 850 – c. 930),
was an Egyptian mathematician.

Scientific Contributions

He is considered the first mathematician to systematically use and accept irrational numbers as solutions and coefficients to equations. His mathematical techniques were later adopted by Fibonacci, thus introducing algebra to Europe.

Abu Kamil made important contributions to algebra and geometry. He was the first mathematician to work easily with algebraic equations with powers higher up to x^8 and solved sets of non-linear simultaneous equations with three unknown variables.

He illustrated the rules of signs for expanding the multiplication. He wrote problems rhetorically, and some of his books did not use any mathematical notation beside those of integers. For example, he uses the Arabic expression "māl shayʾ" ("square-square-thing"). One notable feature of his works is to enumerate all the possible solutions to a given equation.

Following are some of the books by Abū Kāmil.

Book of Algebra (Kitab fi al-jabr wa al-muqabala) is perhaps Abu Kamil's most influential work, which he intended to supersede and expand upon that of Al-Khwarizmi. In this book Abu Kamil solves systems of equations whose solutions are whole numbers and fractions, and accepted irrational numbers (in the form of a square root or fourth root) as solutions and coefficients to quadratic equations.

In Europe, similar material to this book is found in the writings of Fibonacci, and in the 15th century the whole work also appeared in a Hebrew translation by Mordekhai Finzi.

Book of Rare Things in the Art of Calculation (Kitab al-taraif fil-hisab) describes a number of systematic procedures for finding integral solutions for indeterminate equations. It is also the earliest known work where solutions are sought to the indeterminate equations. Abu Kamil also explains certain special methods, such as one problem for which he found 2,678 solutions.

On the Pentagon and Decagon (Kitab al-mukhammas wa'al-mu'ashshar) uses algebraic methods to solve geometrical problems. Abu Kamil uses the equation

$$x^4+3125 = 125x^2$$

to calculate a numerical approximation for the side of a regular pentagon in a circle of diameter 10. He also uses the golden ratio in some of his calculations.

Book of Birds (Kitab al-tair) is a small treatise teaching how to solve indeterminate linear systems with positive integral solutions.

The title is derived from a type of problems known in the east which involve the purchase of different species of birds. Abu Kamil wrote in the introduction:

I found myself before a problem that I solved and for which I discovered a great many solutions; looking deeper for its solutions, I obtained two thousand six hundred and seventy-six correct ones. My astonishment about that was great, but I found out that, when I recounted this discovery, those who did not know me were arrogant, shocked, and suspicious of me. I thus decided to write a book on this kind of calculations, with the purpose of facilitating its treatment and making it more accessible.

On Measurement and Geometry (Kitab al-misaha wa al-handasa) is a manual for non-mathematicians, like land surveyors and other government officials, which presents a set of rules for calculating the volume and surface area of solids (mainly rectangular parallelepipeds, right circular prisms, square pyramids, and circular cones). The first few chapters contain rules for determining the area, diagonal, perimeter, and other parameters for different types of triangles, rectangles and squares.

Ibn al-Nadim in his Fihrist listed his works, including the following:

- Book of Fortune (Kitāb al-falāḥ),
- Book of the Key to Fortune (Kitāb miftāḥ al-falāḥ),
- Book of the Adequate (Kitāb al-kifāya),

- Book of the Kernel (Kitāb al-ʿasīr).
- Book of the Two Errors (Kitāb al-khaṭaʾayn). (A treatise on the use of double false position, known as the.)
- Book on Augmentation and Diminution (Kitāb al-jamʿ wa al-tafrīq), (which gained more attention after historian Franz Woepcke linked it with an anonymous Latin work, Liber augmenti et diminutionis.)
- Book of Estate Sharing using Algebra (Kitāb al-waṣāyā bi al-jabr wa al-muqābala), (which contains algebraic solutions for problems of Islamic inheritance and discusses the opinions of known jurists.)

The works of Abu Kamil influenced other mathematicians, like al-Karaji and Fibonacci, and as such had a lasting impact on the development of algebra.

Many of his examples and algebraic techniques were later copied by Fibonacci in his Practica geometriae and other works. Unmistakable borrowings, but without Abu Kamil being explicitly mentioned.

Biographical Summary

Almost nothing is known about the life and career of Abu Kamil except that he was a successor of al-Khwarizmi, whom he never personally met. Abū Kāmil was born in AD 850 and he died in 930.

17. Al-Hamdani

Abu Muhammad al-Ḥasan ibn Aḥmad ibn Yaqub al-Hamdani

(Arabic: أبو محمد الحسن بن أحمد بن يعقوب الهمداني),

(c. 893-945 A.D),

was a polymath scholar: geographer, chemist, poet, grammarian, historian, and astronomer.

He was from the tribe of Banu Hamdan, Yemen.

Scientific Contributions

In his book "Al-Jawharatayn al-'atīqatayn" Al-Hamdani describes metals (such as gold, silver, and steel), including their physical and chemical properties as well as treatment and processing.

He is also considered one of the earliest scientists who explained gravity of Earth in a way similar to magnetic field behavior.

The work of Al-Hamdani was the subject of extensive 19th-century Austrian research. This resulted in extensive research and publication series. The research was by the Austrian Arabist, Eduard Glaser.

Following publications are derived from Al-Hamdani's works.

Sifat Jazirat ul-Arab (Geography of the Arabian Peninsula) is an important work on the subject of Geography of the Arabian Peninsula.

This manuscript of Al-Hamdani was used by Austrian orientalist, Aloys Sprenger in his publications "Post- und Reiserouten des Orients" (Leipzig, 1864), and "Alte Geographie Arabiens" (Bern, 1875).

It was also used in an edited work by D.H. Müller (Leiden, 1884).

It has been archived (in Arabic) as "worldcat ṣifat ğazīrat al-ʿarab", vol. 12, Leiden, p. 107, 13–14; 149, 17; 154, 3.

Kitāb al-Iklīl min akhbār al-Yaman wa-ansāb Ḥimyar (الإكليل من أخبار اليمن وأنساب حمير) (Crowns from the Accounts of al-Yemen and the genealogies of Ḥimyar). It consists of ten volumes. However, only four volumes have been found (Vol.1, Vol.2, Vol.8 and Vol.10); the other volumes are missing. It describes wars waged by their kings.

Volume 8, is on the citadels and castles of southern Arabia. It has been translated into German, edited and annotated by D.H. Müller and published as "Die Burgen und Schlösser Sudarabiens" (Vienna, 1881).

Other works by al-Hamdani are listed in G. L. Flügel's "Die grammatischen Schulen der Araber" (Leipzig, 1862), pp. 220–221. These include: **History of Saba** and **Language of Himyar and Najran**.

Biographical Summary

The biographical details of al-Hamdani's life are scant, despite his extensive scientific work. He was held in high repute as a grammarian, wrote much poetry, compiled astronomical tables, and is said to have devoted most of his life to the study of the ancient history and geography of Arabia.

Before he was born his family had lived in al-Marashi (المراشي). Then they moved to Sana'a (صنعاء), where al-Hamdani was born in the year 893 AD. His father had been a traveler and had visited Kufa, Baghdad, Basra, Oman and Egypt. At around the age of seven, he started to talk about his desire to travel. Somewhat later he left for Mecca, where he remained and studied for more than six years. Then he departed for Sa'dah (صعدة), where he gathered information on Khawlaan (خولان). Later, he went back to Sanaa and became interested in the land that was Himyar (حمْير), but was imprisoned for two years due to his political views. After his release from prison, he went to Raydah (ريدة) to live under the protection of his own tribe. He compiled most of his books while there and stayed on until his death in 945 AD.

18. Al-ʻIjliyyah

Al-ʻIjliyyah bint al-ʻIjliyy

(Arabic: العجلية بنت العجلي),

was a woman scientist in 10th-century. She was skilled at making astrolabes that were used in Aleppo, Syria.

She is sometimes known in modern popular literature as Mariam al-Asṭurlābiyya (Arabic: مريم الأسطرلابي) but her supposed first name 'Mariam' is not mentioned in the only available source.

Scientific Contributions

She was skilled in the manufacture of astrolabes.

Al-ʻIjliyyah manufactured astrolabes, an astronomical instrument. She was employed by the first Emir of Aleppo, Sayf al-Dawla, who reigned from 944 to 967.

Her father and her were apprentices (tilmīthah) of an astrolabe maker, Nasṭūlus from Baghdad.

The main-belt asteroid 7060 Al-'Ijliya, discovered at Palomar Observatory in 1990, was named in her honor. The naming citation was published on 14 November 2016 (M.P.C. 102252).

Muslims have no taboo for women doing science, and there is no conflict between science and religion. She inspired a character in 2015 award-winning book Binti.

Biographical Summary

According to ibn al-Nadim's Fehrist she was the daughter of an astrolabe maker known as al-'Ijliyy;

19. Al-Saidanani

ʿAbd Allāh ibn al-Ḥasan al-Ḥāsib

was an astronomer and mathematician who lived in the first half of the 10th century.

Scientific Contributions

Ibn an-Nadīm lists the following titles by him:

- *Kitāb fī Ṣunūf aḍ-ḍarb wa-l-qisma* ("Book on the Art of Multiplication and Division")
- *Šarḥ kitāb Muḥammad ibn Mūsā al-Ḫwārizmī fī l-ğabr* ("Commentary on the Book of Muḥammad ibn Mūsā al-Ḫwārizmī on Algebra")
- *Šarḥ kitāb Muḥammad ibn Mūsā al-Ḫwārizmī fī al-ğamʿ wa-t-tafrīq* ("Commentary on the Book of Muḥammad ibn Mūsā al-Ḫwārizmī on Addition and Subtraction")

Biographical Summary

Not much is known about Al-Ṣaidanānī other than that he was an astronomer and mathematician who lived in the first half of the 10th century.

20. Ibn Wahshiyya

Abū Bakr Aḥmad ibn ʿAlī Ibn Waḥshiyyah,

Arabic: أبو بكر أحمد بن علي ابن وحشية),

(died c. 930),

was a Nabataean agriculturalist, toxicologist, and alchemist born in Qussīn, near Kufa in Iraq.

Scientific Contributions

Ibn Waḥshiyyah is the author of the book "Kitāb al-Filāḥa al-Nabaṭiyya" an influential Arabic work on agriculture.

Ibn Wahshiyya's works were written down and redacted after his death by his student and scribe Abū Ṭālib al-Zayyāt. They were used by later agriculturalists.

Ibn al-Nadim, in his Kitāb al-Fihrist (c. 987), lists twenty works attributed to Ibn Wahshiyya.

Many works are attributed to Ibn Wahshiyya, correctly or falsely, including the following:

- Work containing several cipher alphabets.
- A treatise on toxicology called the Book of Poisons, which combines contemporary knowledge on pharmacology.
- Kitāb Shawq al-mustahām fī maʿrifat rumūz al-aqlām ("The Book of the Desire of the Maddened Lover for the Knowledge of Secret Scripts", c. 985). Though it is assessed to be a

spurious attribution to Ibn Wahshiyya, it has been claimed by Egyptologist Okasha El-Daly to have correctly identified the phonetic value of a number of Egyptian hieroglyphs.

Biographical Summary

Ibn Wahshiyya was born in Qussīn, near Kufa in Iraq, and he died 930 AD.

21. Al-Qabisi

Abu al-Saqr Abd al-Aziz ibn Uthman ibn Ali al-Qabisi,

(Arabic: 'Abd al-Azîz, عبدالعزيز القبيصي),

(died 967),

was an astronomer and a mathematician.

Scientific Contributions

Al-Qabisi wrote a modest book on arithmetic, "Risala fi anwâ' al-'adad" (Treatise on the kinds of numbers), in which he discusses perfect numbers and how to form them, and Thābit ibn Qurra's theorem on amicable numbers.

Other works include:

- Risala fi al-ab'âd wa-'l-ajrâm (treatise on distances and bodies);
- Kitāb fi ithbāt ṣinā'at Aḥkām al-nujūm (On Confirming the Art of Astrology);
- Hal al-Zîjat (Solving astronomical tables);
- Risāla fī imtiḥān al-munajjimīn (A treatise for the examination of astrologers)
- Shukūk al-Majisṭī (Doubts on the Almagest);

Biographical Summary

Originally from Qabisa in Iraq, Al-Qabisi later went to Aleppo where he worked for and lived in the palace of Sayf al-Dawla. He died in 967 AD.

22. Al-Uqlidisi

Abu'l Hasan Ahmad ibn Ibrahim Al-Uqlidisi

(Arabic: أبو الحسن أحمد بن ابراهيم الإقليدسي),

(920–980),

was a mathematician, who was working in Damascus and Baghdad.

Scientific Contributions

He wrote the earliest surviving book on the positional use of the Arabic numerals, Kitab al-Fusul fi al-Hisab al-Hindi (The Arithmetic of India) around 952. It is especially notable for its treatment of decimal fractions, and that it showed how to carry out calculations without deletions.

Al-Kashi (d. 1436/7) who treated decimal fractions in his "Miftah al-Hisab", was regarded as the inventor of decimal math. However, al-Uqlidisi, who lived five centuries earlier, is the first mathematician to write about decimal fractions.

Biographical Summary

Not much is known about Uqlidisi other than that he worked in Damascus and Baghdad. He was born in 920 AD and died in 980.

23. Ibn al-A'lam

'Alī ibn al-Ḥusayn Abū l-Qasim al-'Alawi al-Sharif al-Husayni

(Arabic: ابن الأعلم الشريف الحسيني),

(Baghdad, died 985 CE),

was a 10th-century astronomer.

Scientific Contributions

Ibn al-A'lam's main work, a Zīj, was named after his patron "al-Zīj al-'Aḍudī", and was alternatively known as "al-Zīj al-Sharīfi" , or "al-Zīj al-Baghdādi". While the work itself is now lost, scholars were able to somehow reconstruct most of it based on later citations and references in Arabic, Persian, and Greek sources.

Ibn Yunus, the renowned Egyptian astronomer of the 10th century, was also a contemporary to Ibn al-A'lam, whose work he praised and regularly cited. Even after Ibn al-A'lam's death, his influence remained for at least three centuries as evidenced by the tables of Maragheh observatory which were largely based on his work and that of Ibn Yunus.

Ibn al-'Alam seems to have been one of the most prominent astronomers of the 10th century, demonstrated by the impact he seems to have had on both Islamic and Byzantine astronomy.

Biographical Summary

Very little is known about Ibn al-A'lam's life, and even his birth date has not been established by historians. From the little that is known about him, he appears to have been active in Baghdad, working under the patronage of its Buyid ruler, 'Adud al-Dawla (978–983).

His Nisba "al-Alawi al-Sharif" indicates that he was a Sharif and a descendant of Ja'far al-Tayyar.

Ibn al-A'lam died in Baghdad in 985 AD.

24. Maslama al-Majriti

Abu al-Qasim Maslama ibn Ahmad al-Majriti

(Arabic: أبو القاسم مسلمة بن أحمد المجريطي: c. 950–1007),

was a chemist, mathematician, economist and Scholar in Islamic Spain.

His full name is Abu 'l-Qāsim Maslama ibn Aḥmad al-Faraḍī al-Ḥāsib al-Majrīṭī al-Qurṭubī al-Andalusī.

Scientific Contributions

Al-Majriti introduced and improved the astronomical tables of Muhammad ibn Musa al-Khwarizmi. He introduced the techniques of surveying and triangulation.

Al-Majriti aided historians by working out tables to convert Persian dates to Hijri years.

According to Said al-Andalusi, he was the best mathematician and astronomer of his time in al-Andalus. He introduced new surveying methods collaborating with his student ibn al-Saffar.

He wrote a book on taxation and the economy of al-Andalus.

Al-Majrīṭī was centuries ahead of the time when he established a futuristic process of scientific interchange and the advent of networks for scientific communication. This was the foundation for the present-day model of a science community.

He built a school of Astronomy and Mathematics and it marked the beginning of organized scientific research in al-Andalus.

Among his students were Ibn al-Saffar, Abu al-Salt and at-Turtushi.

Biographical Summary

Al-Majrīṭī was born in 950 AD and died in 1007. He was active during the reign of Al-Hakam II.

Not much is known about his life. However, latter day Western writers state that they know about a daughter of his.

25. Ibn Yunus

Abu al-Hasan 'Ali ibn 'Abd al-Rahman ibn Ahmad ibn Yunus al-Sadafi al-Misri

(Arabic: ابن يونس),

(c. 950 – 1009),

was an Egyptian astronomer and mathematician.

Scientific Contributions

Ibn Yunus's research is based on attention to detail and calculations. His most famous work is al-Zij al-Kabir al-Hakimi (c. 1000). It was a handbook of astronomical tables which contained very accurate observations, obtained using very large astronomical instruments.

Delambre noted in his 1819 translation of the Hakemite tables that two of Ibn Yunus' methods for determining the time from solar or stellar altitude were equivalent to the trigonometric identity

$$2 \cos(a) \cos(b) = \cos(a+b) + \cos(a-b)$$

described in Johannes Werner's 16th-century manuscript on conic sections. Although the formulation is now recognized as one of Werner's formulas, it was already known to Ibn Yunus five centuries before, and he used this result in his astronomy work.

Ibn Yunus described 40 planetary conjunctions and 30 lunar eclipses. For example, he accurately describes the planetary conjunction that occurred in the year 1000 as follows:

A conjunction of Venus and Mercury in Gemini, observed in the western sky: The two planets were in conjunction after sunset on the night of Sunday 19 May 1000. The time was approximately eight equinoctial hours after midday on Sunday. Mercury was north of Venus and their latitude difference was a third of a degree.

Modern knowledge of the positions of the planets confirms that his description and his calculation of the distance being one-third of a degree is exactly correct. Ibn Yunus's observations on conjunctions and eclipses were used in Richard Dunthorne and Simon Newcombs' respective calculations of the secular acceleration of the moon.

The crater Ibn Yunus on the Moon is named after him.

Biographical Summary

Ibn Yunus was born in Egypt between 950 and 952. He came from a respected family in Fustat (Cairo).

His father was a historian, biographer, and scholar of hadith, who wrote two volumes about the history of Egypt — one about the Egyptians and one based on traveler commentary on Egypt. A prolific writer, Ibn Yunus' father has been described as "Egypt's most celebrated early historian and first known compiler of a biographical dictionary devoted exclusively to Egyptians".

His great-grandfather had been an associate of the noted legal scholar Imam Shafi.

Early in the life of Ibn Yunus, the Fatimid dynasty came to power and the new city of Cairo was founded. In Cairo, he worked as an astronomer for the Fatimid dynasty for twenty-six years, first for the Caliph al-Aziz and then for al-Hakim. Ibn Yunus dedicated his most famous astronomical work, *al-Zij al-Kabir al-Hakimi*, to the latter.

As well as for his mathematics, Ibn Yunus was also known as a poet.

26. Ibn al-Saffar

Ibn al-Saffar

(Arabic: ابن الصَّفَّار),

(born in Cordoba, died in the year 1035 at Denia),

was an astronomer in Al-Andalus. He was the son of a brass worker, and his full name was Abu al-Qasim Ahmad ibn Abd Allah ibn Umar al-Ghafiqī ibn al-Saffar al-Andalusi.

Scientific Contributions

He worked at the school founded by his teacher, Al-Majriti, in Córdoba. He wrote a treatise on the astrolabe, a text that was in active use until the 15th century.

Kepler's work was based on this treatise. However, the so called "Laws of Kepler" bear the name of Kepler only. There is an abundance of such practices among the Western writers. For example, Paul Kunitzsch argued that a Latin treatise on the astrolabe long attributed to Mashallah, and used by Chaucer to write A Treatise on the Astrolabe, is in fact written by Ibn al-Saffar.

He also wrote a commentary on the Zij al-Sindhind.

Ibn al-Saffar measured the coordinates of Mecca.

The exoplanet Saffar, also known as Upsilon Andromedae b, is named in his honor. And Saffar Island in Antarctica is also named to honor Ibn al-Saffar.

Biographical Summary

Ibn al-Saffar was born in Cordoba. He was the son of a brass worker and a student of Al-Majriti. He died in the year 1035 at Denia.

27. Ibn al-Samh

Abū al-Qāsim Aṣbagh ibn Muḥammad ibn al-Samḥ al-Gharnāṭī al-Mahri

(Arabic: أصبغ المهري),

(born 979, Córdoba; died 1035, Granada),

was a mathematician and astronomer from Al-Andalus. He is also known as Ibn al-Samḥ.

Scientific Contributions

Ibn al-Samḥ was an astronomer at the university founded by Al-Majriti in Córdoba. Political unrest forced him to move to Granada, where he worked with Ḥabbūs ibn Māksan.

He wrote a treatises on the construction and use of the astrolabe.

Ibn al-Samḥ researched the first known work on the planetary equatorium, an analogue computer based on a mechanic-graphical device.

In mathematics he is remembered for a commentary on Euclid and for contributions to early algebra, among other works.

The Western writers in their lack of understanding have invented a fictitious writer called "Abulcasim". Ibn al-Samḥ is one of several writers represented by this fictitious name.

The exoplanet Samh, also known as Upsilon Andromedae c, is named in the honor of Ibn al-Samḥ, as part of the IAU's NameExoWorlds project.

Biographical Summary

Ibn al-Samḥ was born in 979 AD in Córdoba. He died in 1035 in Granada.

28. Abi al-Rijal

Abū al-Ḥasan 'Alī ibn Abī al-Rijāl al-Shaybani (Arabic: أبو الحسن علي ابن أبي الرجال), also known as *Haly Abenragel,* was an astronomer of the late 10th and early 11th century.

Scientific Contributions

Abī al-Rijāl wrote the book: Kitāb al-bāri' fī aḥkām an-nujūm. However, only its European translations survived.

- Kitāb al-bāri' fī aḥkām an-nujūm was translated by Yehudā ben Moshe into Old Castilian for Alfonso X of Castile in 1254 under the title El libro conplido en los iudizios de las estrellas ("The complete book on the judgment of the stars").
- The only surviving manuscript of the Old Castilian translation is MS 3605 at the National Library in Madrid, which however only contains 5 of the 8 books of the complete Old Castilian translation.
- In 1485 at Venice a complete copy of the Old Castilian manuscript was translated into Latin and published by Erhard Ratdolt as: Praeclarissimus liber completus in judiciis astrorum ("The very famous complete book on the judgment of the stars"). This printing (and later Latin versions) is commonly known as De iudiciis astrorum (or De judiciis astrorum).

- An edition created in 1523 in Venice and presented in Latin, is held in the Qatar National Library.
- His Tractatus de cometarum significationibus per xii signa zodiaci (Treatise on the Significations of Comets in the twelve Signs of the Zodiac) was printed in Nürnberg in 1563 as an addendum to Marcus Frytsch's Meteororum.

Biographical Summary

Abī al-Rijāl was a court astronomer to the Tunisian prince al-Mu'izz ibn Bâdis in the first half of the 11th century. He died after 1037 in Kairouan in what is now Tunisia.

The practice of the European writers to change the names of the Arabian writings and their authors has contributed to a loss of biographical information, and the integrity of the research contents, as well as proper attribution of the research credits.

For example, his name was Abū al-Ḥasan 'Alī ibn Abī al-Rijāl al-Shaybani (Arabic: أبو الحسن علي ابن أبي الرجال) but he is called by varied names that depart from the original name drastically. The European writers have called him by a lot of names, for example, *Haly*, *Hali*, and *Albohazen Haly filii Abenragel* or *Haly Abenragel*. This kind of disregard and arbitrariness for original sources is very detrimental to academic integrity of research contributions and traditions.

29. Ibn al-Haytham

Abū ʿAlī al-Ḥasan ibn al-Ḥasan ibn al-Haytham

(Arabic: أبو علي، الحسن بن الحسن بن الهيثم),

(c. 965 – c. 1040),

was a mathematician, astronomer, and physicist. He is also known as Ibn Al-Haytham.

Scientific Contributions

Ibn Al-Haytham is regarded as the father of modern optics, and someone who established a true scientific approach.

Ibn Al-Haytham was a trail blazer in the principles of optics and visual perception in particular. His most influential work is titled Kitāb al-Manāẓir (Arabic: كتاب المناظر, "Book of Optics"), written during 1011–1021, which survived in a Latin edition.

He was a polymath, and also wrote on philosophy, theology and medicine.

Ibn al-Haytham was the first to explain that vision occurs when light reflects from an object and then passes to one's eyes. He was also the first to demonstrate that vision occurs in the brain, rather than in the eyes.

Prior to that there was the emission theory. It was supported by such thinkers as Euclid and Ptolemy, who believed that sight worked by the eye emitting rays of light.

Another theory was the intromission theory supported by Aristotle and his followers. According to this theory, physical forms entered the eye from an object.

Ibn Al-Haytham was World's True Scientist, being a pioneer in the scientific methodology, and a proponent of the concept that

a hypothesis must be supported by experiments based on confirmable procedures or mathematical evidence.

Five centuries before Renaissance in Europe the Muslim scientists were already there.

Ibn Al-Haytham's most famous work is his seven-volume treatise on optics "Kitab al-Manazir (Book of Optics)", written from 1011 to 1021.

Only some European translations have survived. Following are some of these translations, and as usual multiple names have been used for the author.

- Optics was translated into Latin by an unknown scholar at the end of the 12th century or the beginning of the 13th century.
- This work enjoyed a great reputation. The Latin version of De aspcctibus was translated at the end of the 14th century into Italian vernacular, under the title De li aspecti.
- It was printed by Friedrich Risner in 1572, with the title "Opticae thesaurus: Ibn Al-Haythami Arabis libri septem, nuncprimum editi; Eiusdem liber De Crepusculis et nubium ascensionibus" (English: Treasury of Optics: seven books by

the Arab Ibn Al-Haytham, first edition; by the same, on twilight and the height of clouds).

- Risner is also the author of the name variant "Ibn Al-Haytham"; before Risner he was known in the west as Alhacen.
- Works by Ibn Al-Haytham on geometric subjects were discovered in the Bibliothèque nationale in Paris in 1834 by E. A. Sedillot.
- In all, A. Mark Smith has accounted for 18 full or near-complete manuscripts, and five fragments, which are preserved in 14 locations, including one in the Bodleian Library at Oxford, and one in the library of Bruges.

Ibn Al-Haytham showed through experimentation that
light travels in straight lines, and carried out various experiments with lenses, mirrors, refraction, and reflection.

His analyses of reflection and refraction considered the vertical and horizontal components of light rays separately.

Ibn Al-Haytham studied
the process of sight, the structure of the eye, image formation in the eye, and the visual system.

Ian P. Howard argued in a 1996 Perception article
that Ibn Al-Haytham should be credited with many discoveries and theories previously attributed to Western Europeans, writing centuries later.

For example,

he described what became in the 19th century Hering's law of equal innervation. Ibn Al-Haytham wrote a description of vertical horopters 600 years before Aguilonius that is actually closer to the modern definition than Aguilonius's.

A copy of Apollonius' Conics, written in Ibn al-Haytham's own handwriting exists in Aya Sofya: (MS Aya Sofya 2762, 307 fob., dated Safar 415 a.h. [1024]), the year of his death. Conics have been pivotal for Newton to describe the orbits of the planets in the solar system.

Following are some additional research achievements of Ibn al-Haytham.

- Besides the Book of Optics, Ibn Al-Haytham wrote several other treatises on the same subject, including his Risala fi l-Daw' (Treatise on Light). He investigated the properties of luminance, the rainbow, eclipses, twilight, and moonlight. Experiments with mirrors and the refractive interfaces between air, water, and glass cubes, hemispheres, and quarter-spheres provided the foundation for his theories.
- Ibn al-Haytham discussed the physics of the celestial region in his Epitome of Astronomy, arguing that Ptolemaic models must be understood in terms of physical objects rather than abstract hypotheses. In other words that it should be possible to create physical models.
- Ibn al-Haytham wrote Maqala fi daw al-qamar (On the Light of the Moon).

- In his work, Ibn Al-Haytham discussed theories on the motion of a body. In his Treatise on “al-makan” (place or space), Ibn al-Haytham disagreed with Aristotle's view that nature abhors a void. He used geometry to demonstrate that space (al-makan) is the imagined three-dimensional void between the inner surfaces of a containing body.
- Ibn Al-Haytham also discussed space perception and its epistemological implications in his Book of Optics.
- In "tying the visual perception of space to prior bodily experience, Ibn Al-Haytham unequivocally rejected the intuitiveness of spatial perception and, therefore, the autonomy of vision. Without tangible notions of distance and size for correlation, sight can tell us next to nothing about such things."
- Ibn Al-Haytham came up with many theories that shattered what was known of reality at the time. These ideas of optics and perspective did not just tie into physical science, rather into existential philosophy. This led to viewpoints being upheld to the point that there is an observer and their perspective, which in this case is reality.
- In his “On the Configuration of the World” Ibn Al-Haytham presented a detailed description of the physical structure of the earth. The book was eventually translated into Hebrew and Latin in the 13th and 14th centuries and subsequently had an

influence on astronomers such as Georg von Peuerbach during the European Middle Ages and Renaissance.

- In his "Al-Shukūk ʿalā Batlamyūs", published at some time between 1025 and 1028, Ibn Al-Haytham criticized Ptolemy's Almagest, Planetary Hypotheses, and Optics, pointing out various contradictions he found in these works, particularly in astronomy.
- Having pointed out the problems, Ibn Al-Haytham appears to have intended to resolve the contradictions he pointed out in Ptolemy in a later work. Ibn Al-Haytham believed there was a "true configuration" of the planets that Ptolemy had failed to grasp. In the "Doubts Concerning Ptolemy" Ibn Al-Haytham set out his views on the difficulty of attaining scientific knowledge and the need to question existing authorities and theories:
  - Truth is sought for itself [but] the truths, [he warns] are immersed in uncertainties [and the scientific authorities (such as Ptolemy, whom he greatly respected) are not immune from error...
  - He held that the criticism of existing theories—which dominated this book—holds a special place in the growth of scientific knowledge.
- Ibn Al-Haytham's "The Model of the Motions of Each and the Seven Planets" was written c. 1038. Only one damaged

manuscript has been found, with only the introduction and the first section on the theory of planetary motion surviving. (There was also a second section on astronomical calculation, and a third section, on astronomical instruments.)

- Following on from his "Doubts on Ptolemy", Ibn Al-Haytham described a new, geometry-based planetary model, describing the motions of the planets in terms of spherical geometry, infinitesimal geometry and trigonometry. He kept a geocentric universe and assumed that celestial motions are uniformly circular, which required the inclusion of epicycles to explain observed motion with departure from circular motion. However, he managed to eliminate Ptolemy's equant. In general, his model didn't try to provide a causal explanation of the motions, but concentrated on providing a complete, geometric description that could explain observed motions without the contradictions inherent in Ptolemy's model.
- Ibn Al-Haytham wrote a total of twenty-five astronomical works,
  - some concerning technical issues such as Exact Determination of the Meridian,
  - a second group concerning accurate astronomical observation,

- a third group concerning various astronomical problems and questions such as the location of the Milky Way.
- Ibn Al-Haytham made the first systematic effort of evaluating the Milky Way's parallax, combining Ptolemy's data and his own. He concluded that the parallax is (probably very much) smaller than Lunar parallax, and the Milky way should be a celestial object. Though he was not the first who argued that the Milky Way does not belong to the atmosphere, he is the first who did quantitative analysis for the claim.
- The fourth group consists of ten works on astronomical theory, including the Doubts and Model of the Motions discussed above.

- In mathematics, Ibn Al-Haytham built on the mathematical works of Euclid and Thabit ibn Qurra and began the work on "the link between algebra and geometry". This laid the foundation of the new discipline of "*analytic geometry*" which is commonly attributed to Descartes.
- He developed a formula for summing the first 100 natural numbers, and proved it using geometry.
- Ibn Al-Haytham explored what is now known as the Euclidean parallel postulate, the fifth postulate in Euclid's Elements,

using a proof by contradiction, and in effect introducing the concept of *motion into geometry*.

- In elementary geometry, Ibn Al-Haytham attempted to solve the problem of squaring the circle using the area of lunes (crescent shapes).
- Generalizing to *non-Euclidean geometry*, Ibn al-Hatham formulated what latter got the name Lambert quadrilateral; and Boris Abramovich Rozenfeld argued to name it the "Ibn al-Haytham–Lambert quadrilateral".
- Ibn Al-Haytham's contributions to number theory include his work on perfect numbers. In his Analysis and Synthesis, he was the first to state that every even perfect number is of the form $2^{n-1}(2^n - 1)$ where $2^n - 1$ is prime. However, he was not able to prove this result, and Euler later proved it in the 18th century. It is now called the *Euclid–Euler theorem*, totally bypassing the work of Ibn Al-Haytham.
- Ibn Al-Haytham solved problems involving congruences using what is now called *Wilson's theorem*. In his Opuscula, Ibn Al-Haytham considers the solution of a system of congruences, and gives two general methods of solution. His first method, the canonical method, involved Wilson's theorem, while his second method involved a version of the Chinese remainder theorem. Please note how Ibn Al-Haytham's research

remains unacknowledged and the work gets attributed entirely to Wilson.

- Ibn Al-Haytham discovered the sum formula for the fourth power, using a method that could be generally used to determine the sum for any integral power. He used this to find the volume of a paraboloid. He could find the integral formula for any polynomial without having developed a general formula.
- Ibn Al-Haytham wrote a Treatise on the "Influence of Melodies" on the Souls of Animals, although no copies have survived. It appears to have been concerned with the question of whether animals could react to music, for example whether a camel would increase or decrease its pace. This research is pertinent to the tradition where a camel driver in front of a caravans, consisting of a long row of camels, sings melodiously.
- In engineering, one account of his career as a civil engineer has him summoned to Egypt by the Fatimid Caliph, Al-Hakim bi-Amr Allah, to regulate the flooding of the Nile River. He carried out a detailed scientific study of the annual inundation of the Nile River, and he drew plans for building a dam, at the site of the modern-day Aswan Dam. His field work, however, later made him aware of the impracticality of this scheme, and he soon feigned madness so he could avoid punishment from the Caliph.

- Ibn Al-Haytham was a Muslim and most sources report that he was a Sunni and a follower of the Ash'ari school. Ziauddin Sardar says that some of the greatest Muslim scientists, such as Ibn al-Haytham and Abū Rayhān al-Bīrūnī, who were pioneers of the scientific method, were themselves followers of the Ashʿari school of Islamic theology. Ashʿarites believed that faith or taqlid should apply only to Islam and not to any ancient Hellenistic authorities. Likewise, Ibn al-Haytham viewed that taqlid should apply only to prophets of Islam and not to any other authorities. This formed the basis for much of Ibn Al-Haytham's scientific skepticism and criticism against Ptolemy and other ancient authorities in his "Doubts Concerning Ptolemy" and "Book of Optics".
- Ibn Al-Haytham wrote a work on Islamic theology in which he discussed prophethood and developed a system of philosophical criteria to discern its false claimants in his time. He also wrote a treatise entitled "Finding the Direction of Qibla by Calculation" in which he discussed finding the Qibla mathematically.

These demonstrate that Ibn Al-Haytham made foundational contributions to optics, number theory, geometry, astronomy and natural philosophy. Ibn Al-Haytham's work is credited with contributing a new emphasis on experiment.

His main work, Kitab al-Manazir (Book of Optics), was known in the Muslim world mainly, but not exclusively. A Latin translation of the Kitab al-Manazir was made probably in the late twelfth or early thirteenth century. This translation was read by and greatly influenced a number of European scientists including

Roger Bacon, Robert Grosseteste, Witelo, Giambattista della Porta, Leonardo da Vinci, Galileo Galilei, Christiaan Huygens, René Descartes, and Johannes Kepler.

However, they reported their research works largely without reference to the original sources, i.e., works of giants like Ibn Al-Haytham.

Ibn Al-Haytham wrote more than 200 works on a wide range of subjects, of which at least 96 of his scientific works are known. Most of his works are now lost, but more than 50 have survived to some extent.

Nearly half of his surviving works are on mathematics, 23 of them are on astronomy; 14 of them are on optics; with a few on other subjects.

Not all his surviving works have yet been studied, but following have been.

1. Book of Optics (كتاب المناظر)
2. *Analysis and Synthesis* (مقالة في التحليل والتركيب)
3. Balance of Wisdom (ميزان الحكمة)
4. Corrections to the Almagest (تصويبات على المجسطي)

5. *Discourse on Place* (مقالة في المكان)
6. Exact Determination of the Pole (التحديد الدقيق للقطب)
7. Exact Determination of the Meridian (رسالة في الشفق)
8. Finding the Direction of Qibla by Calculation (كيفية حساب اتجاه القبلة)
9. Horizontal Sundials (المزولة الأفقية)
10. *Hour Lines* (خطوط الساعة)
11. Doubts Concerning Ptolemy (شكوك على بطليموس)
12. Maqala fi'l-Qarastun (مقالة في قرسطون)
13. On Completion of the Conics (إكمال المخاريط)
14. On Seeing the Stars (رؤية الكواكب)
15. *On Squaring the Circle* (مقالة فى تربيع الدائرة)
16. *On the Burning Sphere* (المرايا المحرقة بالدوائر)
17. On the Configuration of the World (تكوين العالم)
18. *On the Form of Eclipse* (مقالة فى صورة الكسوف)
19. *On the Light of Stars* (مقالة في ضوء النجوم)
20. On the Light of the Moon (مقالة في ضوء القمر)
21. *On the Milky Way* (مقالة في درب التبانة)
22. On the Nature of Shadows (كيفيات الإظلال)
23. On the Rainbow and Halo (مقالة في قوس قزح)
24. *Opuscula* (Minor Works)
25. Resolution of Doubts Concerning the Almagest (تحليل شكوك حول الجست)
26. Resolution of Doubts Concerning the Winding Motion

27. The Correction of the Operations in Astronomy (تصحيح العمليات في الفلك)
28. The Different Heights of the Planets (اختلاف ارتفاع الكواكب)
29. The Direction of Mecca (اتجاه القبلة)
30. The Model of the Motions of Each of the Seven Planets (نماذج حركات الكواكب السبعة)
31. The Model of the Universe (نموذج الكون)
32. The Motion of the Moon (حركة القمر)
33. The Ratios of Hourly Arcs to their Heights
34. *The Winding Motion* (الحركة المتعرجة)
35. *Treatise on Light* (رسالة في الضوء)
36. *Treatise on Place* (رسالة في المكان)
37. Treatise on the Influence of Melodies on the Souls of Animals (تأثير اللحون الموسيقية في النفوس الحيوانية)
38. كتاب في تحليل المسائل الهندسية (A book in engineering analysis)
39. الجامع في أصول الحساب (The whole in the assets of the account)
40. قول فى مساحة الكرة (Say in the sphere)
41. القول المعروف بالغريب فى حساب المعاملات (Saying the unknown in the calculation of transactions)
42. خواص المثلث من جهة العمود (Triangle properties from the side of the column)
43. رسالة فى مساحة المسجم المكافى (A message in the free space)
44. شرح أصول إقليدس (Explain the origins of Euclid)
45. المرايا المحرقة بالقطوع (The burning mirrors of the rainbow)

Biographical Summary

Ibn al-Haytham (Ibn Al-Haytham) was born in c. 965 in Basra, Iraq. His initial education was in the study of religion and service to the community, as is prevalent in Muslim civilization. Next, he delved into the study of mathematics and science. He held a position with the title vizier in his native Basra, and made a name for himself for his knowledge of applied mathematics.

Ibn al-Haytham was invited to Fatimid Caliph by al-Hakim in order to realize a hydraulic project at Aswan. However, Ibn al-Haytham was forced to concede the impracticability of this project. Upon his return to Cairo, he was given an administrative post. After he proved unable to fulfill this task as well, he contracted the ire of the caliph Al-Hakim bi-Amr Allah, and is said to have been forced into hiding until the caliph's death in 1021. After that his confiscated possessions were returned to him.

Legend has it that Ibn al-Haytham feigned madness and was kept under house arrest during this period. While under house arrest, Ibn al-Haytham wrote his influential "Book of Optics". He continued to live in Cairo, in the neighborhood of the famous University of al-Azhar, and lived from the proceeds of his writings until his death in c. 1040.

Among his students were Sorkhab (Sohrab), a Persian from Semnan, and Abu al-Wafa Mubashir ibn Fatek, an Egyptian prince.

Ibn Al-Haytham argued that the duty of the man who investigates the writings of scientists, if learning the truth is his goal, is to make himself an enemy of all that he reads, and ... attack it from every side. He should also suspect himself as he performs his critical examination of it, so that he may avoid falling into either prejudice or leniency.

There are occasional references to theology or religious sentiment in his technical works, e.g., in "Doubts Concerning Ptolemy" he asserts that truth is sought for its own sake ... Finding the truth is difficult, and the road to it is rough. For the truths are plunged in obscurity. ... God, however, has not preserved the scientist from error and has not safeguarded science from shortcomings and faults. If this had been the case, scientists would not have disagreed upon any point of science...

In "The Winding Motion" he asserts that from the statements made by the noble Shaykh, it is clear that he believes in Ptolemy's words in everything he says, without relying on a demonstration or calling on a proof, but by pure imitation (taqlid); that is how experts in the prophetic tradition have faith in Prophets, may the blessing of God be upon them. But it is not the way that mathematicians have faith in specialists in the demonstrative sciences.

Regarding the relation of objective truth and God he asserts that I constantly sought knowledge and truth, and it became my belief that for gaining access to the effulgence and closeness to God, **there is no better way than that of searching for truth and knowledge.**

30. Ali ibn Khalaf

Alī ibn Khalaf

(Arabic: علي بن خلف الأندلسي),

was an Andalusian astronomer who belonged to the scientific circle of Ṣāʿid al- Andalusī.

Scientific Contributions

Alī ibn Khalaf devised, with help from al-Zarqali, the universal astrolabe.

The European version of it, Libros del Saber (1227) of Alfonso X of Castile, contained both Khalaf and al-Zarqali's designs.

Biographical Summary

Alī ibn Khalaf was an Andalusian astronomer of eleventh century AD who belonged to the scientific circle of Ṣāʿid al- Andalusī.

31. Ibn Khalaf al-Muradi

Ibn Khalaf al-Murādī

(Arabic: أبو جعفر علي ابن خلف المرادي),

was an Andalusian engineer in 11th century.

Scientific Contributions

Al-Murādī was the author of the technological manuscript entitled Kitāb al-asrār fī natā'ij al-afkār (Arabic: كتاب الأسرار في نتائج الأفكار, The Book of Secrets in the Results of Thoughts or The Book of Secrets in the Results of Ideas).

It was copied and used at the court of Alfonso VI of León and Castile in Christian Spain in the 11th century.

The manuscript includes information about a "Castle and Gazelle Clock" and many other forms of complicated clocks and devices.

In 2008, the Book of Secrets of al-Muradi was published in facsimile, translated in English/Italian/French/Arabic and in an electronic edition with all machines interpreted in 3D, by the Italian study center Leonardo.

Biographical Summary

Ibn Khalaf al-Murādī was an Andalusian engineer in 11th century AD.

32. Al-Biruni

Abu Rayhan Muhammad ibn Ahmad al-Biruni

(973 – after 1050), commonly known as al-Biruni,

was a Khwarazmian scholar and polymath from Iran.

Scientific Contributions

Al-Biruni has been called variously the "founder of Indology", "Father of Comparative Religion", "Father of modern geodesy", and the first anthropologist.

Al-Biruni was well versed in physics, mathematics, astronomy, and natural sciences, and also distinguished himself as a historian, chronologist, and linguist. He studied almost all the sciences of his day and researched tirelessly in many fields of knowledge.

Astronomical, geographical, mathematical, and engineering research possessed not only a scientific but also a religious dimension. The Caliphs, therefore, supported learning and research. In Islam worship and prayer require a knowledge of the direction of Mecca, knowledge of Lunar calendar, Miqat boundaries for Haj and Umra, calculations for heritage apportionments, water management for Wudu, and mecca-directed architecture for mosques. This requires the use of astronomical data, geography, geometry, spherical geometry, trigonometry, and instrumentation.

Ninety-five of 146 books known to have been written by Al-Bīrūnī are devoted to astronomy, mathematics, and related subjects like mathematical geography. He lived during the Islamic Abbasid Caliphs who patronized knowledge and learning. In carrying out his research, Al-Biruni used a variety of different techniques dependent upon the particular field of study involved.

His major work is an astronomical and mathematical text. He states:

"I have begun with Geometry and proceeded to Arithmetic and the Science of Numbers, and finally to astronomy and the structure of the Universe; for no one is worthy of the style and title of an Astronomer who is not thoroughly conversant with these four sciences."

He was the first to make the semantic distinction between astronomy and astrology and, in a later work, wrote a refutation of astrology, in contradistinction to the legitimate science of astronomy, for which he expresses wholehearted support.

He wrote an extensive commentary "Taḥqīq mā li-l-Hind" on Indian astronomy, ad he resolved the matter of Earth's rotation in a work on astronomy that is no longer extant, namely "Miftah-ilm-alhai'a" (Key to Astronomy).

In his description of Sijzi's astrolabe he hints at contemporary debates about the movement of the earth. He carried on a lengthy correspondence and sometimes heated debate with Ibn Sina, in which

Biruni attacks Aristotle's celestial physics: he argues by simple experiments that the vacuum state must exist. Also, he is "amazed" by the weakness of Aristotle's argument against elliptical orbits on the basis that they would create a vacuum; and he attacks the immutability of the celestial spheres.

In his major astronomical work, the Mas'ud Canon, Biruni observed that, contrary to Ptolemy, the sun's apogee (highest point in the heavens) was mobile, not fixed. He wrote a treatise on the astrolabe, describing how to use it to tell the time and as a quadrant for surveying. One particular diagram of an eight geared device, that Al-Biruni presented, could be considered an ancestor of later astrolabes and clocks. Al-Biruni's eclipse data was used by Dunthorne in 1749 to help determine the acceleration of the moon, and his data on equinox times and eclipses was used as part of a study of Earth's past rotation.

Al-Biruni was the person who first, in 1000 AD, subdivided the sexagesimal hour into minutes, seconds, thirds and fourths.

Biographical Summary

The name of al-Biruni is derived from the Persian word bīrūn (meaning 'outskirts'). Al-Biruni was born in 973 AD in the Bīrūn district, the outer district (Bīrūn) of Kath, the capital of the Afrighid dynasty of Khwarezm in Central Asia – now part of the autonomous republic of Karakalpakstan in the northwest of Uzbekistan.

Al-Biruni spent the first twenty-five years of his life in Khwarezm where he studied Islamic jurisprudence, theology, grammar, mathematics, astronomy, medicine and philosophy and dabbled not only in the field of physics, but also in those of most of the other sciences. He was sympathetic to the Afrighids, who were overthrown by the rival dynasty of Ma'munids in 995. He left his homeland for Bukhara; there he corresponded with Avicenna and there are extant exchanges of views between these two scientists.

In 998, he went to the court of the Ziyarid amir of Tabaristan, Qabus (r. 977–981, 997–1012). There he wrote his first important work, al-Athar al-Baqqiya 'an al-Qorun al-Khaliyya (literally: "The remaining traces of past centuries" and translated as "Chronology of ancient nations" or "Vestiges of the Past") on historical and scientific chronology, probably around 1000 C.E., though he later made some amendments to the book.

He also visited the court of the Bavandid ruler Al-Marzuban. Accepting the definite demise of the Afrighids at the hands of the Ma'munids, he made peace with the latter who then ruled Khwarezm; their court at Gorganj (also in Khwarezm) was gaining fame for its gathering of brilliant scientists.

In 1017, Mahmud of Ghazni took Rey. Most scholars, including al-Biruni, were taken to Ghazni, the capital of the Ghaznavid dynasty. Biruni was made court astrologer and accompanied Mahmud on his campaigns into India, living there for a few years; he was forty-four years old then. Biruni became acquainted with India, and he wrote his study of India, finishing it around 1030. Along with his writing, Al-Biruni also made sure to extend his study to science while on the expeditions. He sought to find a method to measure the height of the sun, and created a makeshift quadrant for that purpose.

Belonging to the Sunni Ash'ari school, al-Biruni nevertheless also associated with Maturidi theologians. He was, however, very critical of the Mu'tazila, particularly criticizing al-Jahiz and Zurqan. He also repudiated Avicenna for his views on the eternality of the universe.

Al Biruni died sometime after 1050 AD.

33. Yusuf al-Mu'taman ibn Hud

Abu Amir Yusuf ibn Ahmad ibn Hud

(Arabic: أبو عامر يوسف إبن أحمد إبن هود), more commonly known as al-Mu'taman Billah (Arabic: المؤتمن بالله),

(died c. 1085),

was a mathematician.

He was also one of the kings of the Taifa of Zaragoza.

Scientific Contributions

The main work of al-Mu'taman was his Book of Perfection (Kitab al-Istikmal). This book was a compendium of the teachings of Thabit ibn Qurra, the Banu Musa and Ibn al-Haytham. More significantly, it contained his original research, including the proofs to his original theorems.

The Kitab al-Istikmal could not be completed when he died in 1085; and the encyclopedist Ibn al-Akfani said that had the Istikmal been completed, it would have made the existing geometrical literature superfluous.

The work was sent to Egypt by Maimonides, and from there it spread to Baghdad in the 14th century.

The Kitab al-Istikmal deals with irrational numbers, conic sections, quadrature of the parabolic segment, volumes and areas of various geometric objects, and the drawing of the tangent to a circle, among other mathematical problems. It includes a chapter for arithmetic, two chapters for geometry and two others for stereometry.

Al-Mu'taman is the author of the first known formulation of a theorem, which can be stated as follows:

Let ABC be a triangle and D, E, F points on the sides BC, CA, and AB. We draw the lines AD, BE and CF. These three lines intersect at one point if and only if

AF/FB = (EA/CE) (DC/BD)

In 1678 Italian geometer Giovanni Ceva reported it without acknowledgment in his book De lineis rectis and is known as Ceva's Theorem.

Biographical Summary

Al-Mu'taman was the third king of the Banu Hud dynasty, reigning from 1081 to 1085, at the height of power of Muslim Zaragoza, following the thriving period of his father Ahmad al-Muqtadir. He continued his father's efforts and created around him a court of intellectuals, living in the beautiful palace of Aljafería.

As king, Al-Mu'taman was a patron of science, philosophy and arts, and was himself a scholar of considerable accomplishment. He was a mathematician, astronomer, and a philosopher. He wrote a significant treatise on mathematics, the Kitab al-Istikmal ("Book of Perfection").

He died in 1085 AD.

Yusuf was born in Zaragoza, in the palace of Aljaferia. When he ascended to the throne on the death of his father in 1081, the taifa of Zaragoza was at its peak.

Al-Muqtadir divided his lands between his two sons: al-Mu'taman received the western part of the taifa with Zaragoza, Tudela, Huesca and Calatayud areas, while al-Mundhir received the coastal zone of the kingdom, including Lérida, Monzón, Tortosa and Dénia.

34. Ali ibn Ridwan

Abu'l Hassan Ali ibn Ridwan Al-Misri

(Arabic: أبو الحسن علي بن رضوان المصري),

(c. 988 - c. 1061),

was a physician, and astronomer.

Scientific Contributions

Ali ibn Ridwan is known for his discovery of the Super Nova SN 1006, which is the brightest astronomical event in recorded history.

The European science community cited him by a fictitious name, Haly or Haly Abenrudian.

Ali ibn Ridwan also researched the theory of induction for inductive reasoning in the construction of proofs.

As a medical Scientist, he was the president of physicians in Egypt.

Biographical Summary

Ali ibn Ridwan was born in Giza, Egypt, in 988 AD and he died in 1061 AD.

35. Said al-Andalusi

Ṣāʿid al-Andalusī

(Arabic: صاعِدُ الأندلسي),

(1029 – July 6, 1070 AD),

was a qadi of Toledo in Muslim Spain, who wrote on the history of science, philosophy and thought.

His full name was Abū al-Qāsim Ṣāʿid ibn Abū al-Walīd Aḥmad ibn Abd al-Raḥmān ibn Muḥammad ibn Ṣāʿid ibn ʿUthmān al-Taghlibi al-Qūrtūbi (صاعِدُ بنُ أحمدَ بن عبد الرحمن بن محمد بن صاعدٍ التَّغْلِبيُّ)).

Scientific Contributions

He was a mathematical scientist with a special interest in astronomy. He authored a famous biographic encyclopedia of science that quickly became popular.

The Ṭabaqāt al-'Umam (Tabaqāt) composed in 1068 is a "history of science" that comprises biographies of the scientists and scientific achievements of eight nations. In the field of nations are the Indians, Persians, Chaldeans, Egyptians, Greeks, Byzantines, Arabs and Jews.

Please note the inconspicuous absence of all the Europe as they were in the dark ages and did not contribute to science and technology.

Ṣāʿid offers an account of the individual contribution each nation makes to the various sciences of arithmetic, astronomy, and medicine, etc. He also includes ancient scientists from Greece, for example, Pythagoras, Socrates, Plato and Aristotle. The second half of the book contains Arab-Islamic contributions to the fields of logic, philosophy, geometry, astronomy, observational methods, calculations in trigonometry and mathematics to determine the length of the year, the eccentricity of the sun's orbit, and the construction of astronomical tables, etc.

The Ṭabaqāt al-'Umam has been transcribed and translated into many different languages in many periods and cultures. The original document is not extant. European writers are generally not careful in translating these researches. As a consequence, there are discrepancies in the translations, including variations in the title of the book; discrepancies in the content of the editions appear with some versions omitting words, sentences, paragraphs or entire sections. Some of the these are presumably because of backwardness of Europe in these fields of science and knowledge. Translators naturally found themselves at loss because of lack of understanding.

Other works of Ṣāʿid al-Andalusī are below.

- *Iṣlāh Ḥarakāt an-Najūm* (اصلاح حركات النجوم) on the corrections of earlier astronomical calculations.
- *Jawāmiʿ akhbār al-umam min al-Arab wa-l Ajam* (جوامع أخبار الأمم من العرب والعجم; 'Universal History of Nations – Arab and Non-Arab').

- *Ṭabaqāt al-'Umam* (طبقات الأمم), a classification of the nations (according to their science and technology achievements).
- An astronomical treatise on Rectification of Planetary Motions and Exposition of Observers' Errors;
- *Maqālāt ahl al-milal wa-l-nihal* (مقالات أهل الملل والنحل; 'Discourses on the Adherents of Sects and Schools').
- *Kitāb al-Qāsī* (كتاب القاصى), 'Book of the Minors'

Biographical Summary

The family of Ṣā'id al-Andalusī , during the civil war, had fled Cordova to take refuge in Almería. Ṣā'id al-Andalusī was born in Almería in 1029 AD. He died in Toledo, in 1070 AD.

His family belonged to the tribe of Taghlib in Arabia.

His grandfather had been qadi (judge) of Sidonia and his father was qadi of Toledo until his death in 1057 when Ṣā'id succeeded him.

Ṣā'id's teachers in Toledo were Abū Muḥammad ibn Hazm (أبو محمد بن حَزْم), Al-Fataḥ ibn al-Qāsim (الفَتْح بن القاسم), and Abū Walīd al-Waqshi (أبو الوليد الوَقّشِي). He was educated in fiqh (law) first in Almería, then Córdoba, before graduating, it seems, in Toledo in 1046, at age eighteen. This information obtains from early Biographers: Ibn Bashkuwāl, Ibn Umaira al-Dhabbi, Al-Safadi and Ahmed Mohammed al-Maqqari.

Toledo was a great center of learning and Ṣā'id studied fiqh (law), tafsir (Qu'ranic exegesis), Arabic language, and al-Adab al-'Arabī (Arabic civilization). His teacher, Abū Isḥaq Ibrāhīm ibn Idrīs al-Tajibī, directed him towards mathematics and astronomy, in which he excelled. On his appointment as qāḍi of Toledo by the governor Yaḥyā al-Qādir, he continued this work and produced several scholarly works, that contributed to the Tables of Toledo.

He taught and directed astronomical research to a group of young scholars, precision-instrument-makers, astronomers and scientists – including the renowned Al-Zarqali – and encouraged them for excellence in research. Their research also contributed to the Tables of Toledo.

36. Al-Jayyani

Abū ʿAbd Allāh Muḥammad ibn Muʿādh al-Jayyānī

(Arabic: أبو عبد الله محمد بن معاذ الجياني),

(989 – 1079),

was a mathematician, Islamic scholar, and a Qadi.

Scientific Contributions

Al-Jayyānī wrote the first known treatise on spherical trigonometry. Al-Jayyānī wrote the book of unknown arcs of a sphere, which is considered the first treatise on spherical trigonometry. Al-Jayyānī's work on spherical trigonometry contains formulae for right-handed triangles, the general law of sines, and the solution of a spherical triangle by means of the polar triangle. This treatise later had a strong influence on European mathematics, and his "definition of ratios as numbers" and "method of solving a spherical triangle when all sides are unknown" are likely to have influenced Regiomontanus.

Biographical Summary

Al-Jayyānī was born in 989 AD, in Andalusian city of Cordova; and he died in 1079, in the Andalusian city of Jaén. Little else is known about his life.

37. Ibn Bassal

Ibn Bassal

(Arabic: ابن بصال),

was an 11th-century Andalusian botanist and agronomist in Toledo and Seville, Spain.

Scientific Contributions

Ibn Bassal's wrote treatises on agronomy and horticulture.

'Kitāb al-Kasd wa 'l-bayān (The Book of Concision and Clarity) was the abridged version of his seminal work "Divan al-filaha". It was abridged during his lifetime, and condensed the Divan al-filaha into a single volume.

Dīwān al-filāḥa (An Anthology of Husbandry), researches a taxonomy of Soils fertility with ten main classes. It was originally a copious manuscript that had been dedicated to the botanical garden of Al-Ma'mūn at Toledo.

Although it had originally been compiled in Arabic, the work was later translated into Castilian in the 13th century, and many years later into Spanish.

Ibn Bassal's systematic book Dīwān al-filāha contains a record of his own research, thus needing no references to other agronomists. He researches over 180 cultivated plants, including:

chickpeas, beans, rice, peas, flax, henbane, sesame, cotton, safflower, saffron, poppies, henna, artichoke; herbs and spices including cumin, caraway, fennel, anise, and coriander; vegetables requiring irrigation or plentiful watering such as cucumbers, melons, mandrake, watermelons, pumpkins and squash, eggplant, asparagus, caper, and colocynth; the root vegetables carrots, radish, garlic, onion, leek, parsnip, the Sudanese pepper, and the dye-yielding madder; leaf vegetables including cabbage, cauliflower, spinach, purslane, amaranth and chard.

He also covers arboriculture, detailing the propagation of the palm, olive, pomegranate, quince, apple, fig, pear, cherry, apricot, plum, peach, almond, walnut, hazelnut, grape, citron, orange, pistachio, pine, cypress, chestnut, holm-oak, deciduous oak, tree of paradise, arbutus, elm and ash.

He researches manure, mixing animal dung with straw, concluding that it is not composed of only one material (animal dung) but is a mixture.

He explains that the sweepings from hot baths included urine and human wastes, and is unsuitable for use as fertilizer unless mixed with other types of manure.

Ibn Bassal gives two recipes for composting pigeon (hamam) and possibly donkey (himar) manure. Bassal says the excessive heat and moist qualities of pigeon dung works well for weaker and less hardy plants, especially those affected by cold temperatures. Human waste, on the other hand, Bassal advises, to use in hot temperatures because there is no heat to it. Pig dung, he cautions, will destroy pastures and poison plants, a view also shared by non-Arab writers like Columella and Cassianus Bassus. Compost made without manure is considered less desirable; Ibn Bassal calls this type muwallid, made with herbage, straw and grass, ashes from ovens, and water. Some of Bassal's text was copied by the Yemeni writers Al-Malik al-Afḍal.

Ibn Bassal's works were studied several centuries later by Abu Jafar Ahmad Ibn Luyūn al-Tujjbi (d.1349) of Almeria who wrote his treatise Kitāb Ibdā' al-malāha wa-inhā' al-rajāha fī usūl sinā'at al-filāha.

Biographical Summary

Ibn Bassal worked at the Abbasid court of Al-Mutamid, for whom he created the Hā'īṭ al-Sulṭān botanical garden in Seville. Originally from Toledo, Ibn Bassal moved to Seville after Alfonso VI conquered Toledo in 1085.

He travelled (on pilgrimage) to the Hejaz, visiting Egypt, Sicily, Syria, and seemingly also countries from Abyssinia and Yemen to Iraq, Persia and India.

38. Al-Zarqali

Abū Isḥāq Ibrāhīm ibn Yaḥyā al-Naqqāsh al-Zarqālī al-Tujibi

(Arabic: إبراهيم بن يحيى الزرقالي),

(1029–1087),

was an astronomer and an instrument designer.

Scientific Contributions

Al-Zarqali's works inspired a generation of astronomers in Al-Andalus, and later, after being translated, were very influential in Europe.

His invention of the Saphaea (a perfected astrolabe) was widely used by navigators until the 16th century.

The crater Arzachel sic.) on the Moon is named after him.

Al-Zarqali corrected geographical data from Ptolemy and Al-Khwarizmi. Specifically, he corrected Ptolemy's estimate of the longitude of the Mediterranean Sea from 62 degrees to the correct value of 42 degrees. In his treatise on the solar year, which survives only in a Hebrew translation, he was the first to demonstrate the motion of the solar apogee relative to the fixed background of the stars. He measured its rate of motion as 12.04 seconds per year, which is remarkably close to the modern calculation of 11.77 seconds. Al-Zarqālī's model for the motion of the Sun, in which the center of the Sun's deferent moved on a small, slowly rotating circle to reproduce the observed motion of the solar apogee, was presented as research findings in the thirteenth century by Bernard of Verdun and in the fifteenth century by Regiomontanus and Peurbach. In the sixteenth century Copernicus employed this model in his De Revolutionibus Orbium Coelestium. However, it is without reference to Al-Zarqali as the original source.

Al-Zarqālī also contributed to the famous Tables of Toledo, an adaptation of earlier astronomical data to the location of Toledo along with the addition of some new material.

Al-Zarqālī was famous as well for his own Book of Tables, of which he compiled many. Al-Zarqālī' almanac contained tables which allowed one to find the days on which the Coptic, Roman, lunar, and Persian months begin, other tables which give the position of planets at any given time, and still others facilitating the prediction of solar and lunar eclipses.

He compiled an almanac that directly provided "the positions of the celestial bodies and needed no further computation". The work provided the true daily positions of the sun for four Julian years from 1088 to 1092, the true positions of five planets every 5 or 10 days over a period of 8 years for Venus, 79 years for Mars, and so forth. There were also other related tables.

His Zij and Almanac were translated into Latin by Gerard of Cremona in the 12th century, and contributed to the birth of a mathematically based astronomy in Christian Europe; and were later incorporated into the Tables of Toledo in the 12th century and the Alfonsine tables in the 13th century.

Al-Zarqālī noted that the path of the center of the primary epicycle is not a circle, as it is for the other planets. Instead, it is approximately oval and similar to the shape of a pignon (or pine nut). Some writers have misinterpreted al-Zarqālī's description of an earth-centered oval path for the center of the planet's epicycle as an anticipation of Johannes Kepler's sun-centered elliptical paths for the planets. Although this may be the first suggestion that a conic section could play a role in astronomy, al-Zarqālī did not apply the ellipse to astronomical theory and neither he nor his Iberian or Maghrebi contemporaries used an elliptical deferent in their astronomical calculations.

Al-Zarqālī wrote two works on the construction of an instrument (an equatorium) for computing the position of the planets. These works were translated into Spanish in the 13th century by order of King Alfonso X in a section of the “Libros del Saber de Astronomia” entitled the "Libros de las laminas de los planetas".

He also invented a perfected kind of astrolabe known as "the tablet of al-Zarqālī" (al-ṣafīḥā al-zarqāliyya), which was known in Europe under the name Saphaea, without attribution to Al-Zarqali.

Publications:

- "Al Amal bi Assahifa Az-Zijia";
- "Attadbir";

- "Al Madkhal fi Ilm Annoujoum";
- "Rissalat fi Tarikat Istikhdam as-Safiha al-Moushtarakah li Jamiâ al-ouroud";
- "Almanac Arzarchel";

His works influenced Ibn Bajjah (Avempace), Ibn Tufail (Abubacer), Ibn Rushd (Averroës), Ibn al-Kammad, Ibn al-Haim al-Ishbili and Nur ad-Din al-Betrugi (Alpetragius) – please note the names were changed unrecognizably!

Ragio Montanous wrote a book in the 15th century on the advantages of the Sahifah al-Zarqalia. In 1530, the German writer Jacob Ziegler wrote a commentary on one of al-Zarqali's works. In his "De Revolutionibus Orbium Coelestium", in the year 1530, Nicolaus Copernicus quotes the works of al-Zarqali and Al-Battani.

Biographical Summary

Al-Zarqālī was born in 1028 in a village near the outskirts of Toledo, the then capital of the Taifa of Toledo. He had to flee when Toledo was occupied in 1085 by the Christian king of Castile Alfonso VI. Al-Zarqālī died in a Moorish refugee camp in 1087 AD.

He was trained as a metalsmith and due to his burr skills, he was nicknamed Al-Nekkach "the engraver".

He was particularly talented in Geometry and Astronomy. He is known to have taught and visited Córdoba on various occasions, and his extensive experience and knowledge eventually made him the foremost astronomer. Al-Zarqālī was also an inventor, and his works helped to put Toledo at the intellectual center of Al-Andalus.

There is a record of an al-Zarqālī who built a water clock, capable of determining the hours of the day and night and indicating the days of the lunar months. According to a report found in al-Zuhrī's Kitāb al-Juʿrāfīyya, his name is given as Abū al-Q̄āsim bin ʿAbd al-Raḥmān, also known as al-Zarqālī.

39. Al-Tighnari

Muhammad ibn Malik al-Tighnari, Al-Tighnari

(Arabic: الطغنري),

(1073–1118),

was an important Agronomist, Botanist, Physician and author.

Scientific Contributions

Al-Tighnari wrote a treatise on agronomy in 12 chapters entitled Zuhrat al-bustān wa-nuzhat al-adhhan (The Flowers of Garden and Intellect) under the sponsorship of the Almoravid prince Tamim, son of Yusuf Ibn Tashufin. Prince Tamim was governor of the province of Granada and patron of Al-Tighnari and other Agronomists and Botanists.

Biographical Summary

Al-Tighnari was born in 1073 AD into a family of noble Arab lineage in Tignar, a village a few kilometers north of Granada. He died in 1118 AD.

Al-Tighnari was supportive of agronomy, food security, and fair taxation.

40. Al-Tughrai

Mu'ayyad al-Din Abu Isma'il al-Husayn ibn Ali ibn Muhammad ibn Abd al-Samad al-Du'ali al-Kināni al-Tughra'i

(Arabic: العميد فخر الكتاب مؤيد الدين أبو إسماعيل الحسين بن علي بن محمد بن عبد الصمد الدؤلي الكناني),

(1061 – 1121),

was a poet and alchemist.

Scientific Contributions

Al-Tughra'i was a well-known and prolific writer on astrology and alchemy, and many of his poems (diwan) are preserved today as well.

In the field of alchemy, al-Tughra'i is best known for his large compendium titled Mafatih al-rahmah wa-masabih al-hikmah, which incorporated extensive extracts from earlier Arabic alchemical writings. Until 1995 these were erroneously attributed to unknown alchemists, due to mistakes and inconsistencies in the transliteration and transcription of his name.

In 1112 AD, al-Tughra'i also composed Kitab Haqa'iq al-istishhad, a rebuttal of a refutation of the occult in alchemy written by Ibn Sina.

Biographical Summary

Mu'ayyad al-Din al-Tughra'i was born in 1061 in Isfahan, Persia. He composed poems in the Arabic language. He was an administrative secretary (therefore the name Tughra'i'). He ultimately became the second-most-senior official (after the vizier) in the civil administration of the Seljuq Empire.

Al-Tughra'i had been appointed vizier to Emir Ghiyat-ul-Din Mas'ud, and upon the death of the emir a power struggle ensued between Mas'ud's sons. Al-Tughra'i sided with the emir's elder son, but the younger prevailed. In retribution, the younger son accused al-Tughra'i of heresy and had him beheaded in 1122.

41. Omar Khayyam

Ghiyāth al-Dīn Abū al-Fatḥ ʿUmar ibn Ibrāhīm Nīsābūrī (Persian: عمر خیّام), commonly known as Omar Khayyam, (18 May 1048 – 4 December 1131), was a Persian polymath, mathematician, astronomer, historian, philosopher, and poet.

He was born in Nishapur, the initial capital of the Seljuk Empire. He was contemporary with the rule of the Seljuk dynasty around the time of the First Crusade.

Scientific Contributions

His pioneering work is on the classification and solution of cubic equations in mathematics, where he provided geometric solutions by the intersection of conics, and similar works in astronomy. However, the Europeans represent him just as a poet, oblivious of his stature as a scientist.

Khayyam was famous during his life as a mathematician. His surviving mathematical works include:

- A commentary on the difficulties concerning the postulates of Euclid's Elements (Risāla fī šarḥ mā aškala min muṣādarāt kitāb Uqlīdis), completed in December 1077,

- On the division of a quadrant of a circle (Risālah fī qismah rub‘ al-dā’irah), undated but completed prior to the treatise on algebra, and
- On proofs for problems concerning Algebra (Maqāla fi l-jabr wa l-muqābala), most likely completed in 1079.
- He, furthermore, wrote a treatise on the binomial theorem and extracting the nth root of natural numbers, which has been lost.

A part of Khayyam's commentary on Euclid's Elements deals with the parallel axiom. The treatise of Khayyam can be considered the first treatment of the axiom based on a more intuitive postulate. Khayyam refutes the previous attempts by other mathematicians to prove the proposition, mainly on grounds that each of them had postulated something that was by no means easier to admit than the Fifth Postulate itself.

He rejects the usage of movement in geometry and therefore dismisses the different attempt by Al-Haytham.

Unsatisfied with the failure of mathematicians to prove Euclid's parallel axiom (also known as Euclid's 5th postulate) from his other postulates, Omar tried to connect the axiom with the Fourth Postulate, which states that all right angles are equal to one another.

Khayyam was the first to consider the three distinct cases of acute, obtuse, and right angle for the summit angles of a quadrilateral, called Khayyam quadrilateral. After proving a number of theorems about them, he showed that Postulate V follows from the right-angle hypothesis (Postulate IV), and refuted the obtuse and acute cases as self-contradictory.

His elaborate attempt to prove the parallel postulate was significant as it clearly shows the possibility of non-Euclidean geometries. The hypotheses of acute, obtuse, and right angles are now known to lead respectively to:

- the non-Euclidean hyperbolic geometry of Gauss-Bolyai-Lobachevsky,
- Riemannian geometry, and
- Euclidean geometry.

Omar Khayyam founded the case for non-Euclidean geometries, however, none of these geometries bear his name.

Tusi's commentaries on Khayyam's treatment of parallels made its way to Europe. John Wallis, professor of geometry at Oxford, translated Tusi's commentary into Latin.

What is generally considered the first step in the eventual development of non-Euclidean geometry is the work of Jesuit geometer Girolamo Saccheri titled "euclides ab omni naevo vindicatus, 1733". However, Saccheri was familiar with the translation by Wallis of Tusi's commentaries on Omar Khayyam's treatment of parallels; and Tusi distinctly states that the work is due to Omar Khayyam.

The American historian of mathematics David Eugene Smith mentions that Saccheri "used the same lemma as the one of Tusi, even lettering the figure in precisely the same way and using the lemma for the same purpose".

The situation is unmistakable enough to postulate that Saccheri's work was borrowed work from Omar Khayyam via Tusi.

Omar Khayyam's treatise on Euclid contains another contribution dealing with the theory of proportions and with the compounding of ratios. Khayyam discusses the relationship between the concept of ratio and the concept of number and explicitly raises various theoretical difficulties. In particular, he contributes to the theoretical study of the concept of an irrational number. Displeased with Euclid's definition of equal ratios, he redefined the concept of a number by the use of a continuous fraction as the means of expressing a ratio.

Rosenfeld and Youschkevitch (1973) argue that "by placing irrational quantities and numbers on the same operational scale, Khayyam began a true revolution in the theory of numbers."

In the Treatise on the Division of a Quadrant of a Circle, Khayyam applied algebra to geometry. In this work, he devoted himself mainly to investigating whether it is possible to divide a circular quadrant into two parts such that the line segments projected from the dividing point to the perpendicular diameters of the circle form a specific ratio. His solution, in turn, employed several curve constructions that led to equations containing cubic and quadratic terms.

Khayyam was the first to conceive a general theory of cubic equations and the first to geometrically solve every type of cubic equation, so far as positive roots are concerned. The treatise on algebra contains his work on cubic equations. It is divided into three parts:

- equations which can be solved with compass and straight edge,
- equations which can be solved by means of conic sections, and
- equations which involve the inverse of the unknown.

Khayyam produced an exhaustive list of all possible equations involving lines, squares, and cubes. He considered three binomial equations, nine trinomial equations, and seven tetranomial equations. For the first- and second-degree polynomials, he provided numerical solutions by geometric construction.

He concluded that there are fourteen different types of cubics that cannot be reduced to an equation of a lesser degree. For these, the construction of his unknown segment with compass and straight edge was not possible, and he proceeded to present geometric solutions to all types of cubic equations using the properties of conic sections.

The prerequisite lemmas for Khayyam's geometrical proof include Euclid VI, Prop 13, and Apollonius II, Prop 12. The positive root of a cubic equation was determined as the abscissa of a point of intersection of two conics, for instance, the intersection of two parabolas, or the intersection of a parabola and a circle, etc.

However, he acknowledged that the arithmetic problem of these cubics was still unsolved, adding that "possibly someone else will come to know it after us". This task remained open until the sixteenth century, where algebraic solution of the cubic equation was found in its generality by Cardano, Del Ferro, and Tartaglia in Renaissance Italy.

According to Omar Khayyam, whoever thinks algebra is a trick in obtaining unknowns has thought it in vain. No attention should be paid to the fact that algebra and geometry are different in appearance. Algebras are geometric facts which are proved by propositions five and six of Book two of Elements. He thus integrated algebra and geometry, the foundations for analytic geometry. Therefore, Rashed and Vahabzadeh have argued that because of his thoroughgoing geometrical approach to algebraic equations, Khayyam laid the foundations for analytic geometry, something that is commonly attributed to Descartes. Europeans have celebrated Khayyam as a poet and ignored his mathematical contributions!

Khayyam wrote that:

“From the Indians one has methods for obtaining square and cube roots, methods based on knowledge of individual cases – namely the knowledge of the squares of the nine digits 12, 22, 32 (etc.) and their respective products, i.e. 2 × 3 etc.

We have written a treatise on the proof of the validity of those methods and that they satisfy the conditions.

In addition, we have increased their types, namely in the form of the determination of the fourth, fifth, sixth roots up to any desired degree.

No one preceded us in this and those proofs are purely arithmetic, founded on the arithmetic of The Elements".

Omar Khayyam wrote a Treatise on Demonstration of Problems of Algebra, in which Khayyam alludes to a book he had written on the extraction of n-th root of the numbers using a law he had discovered which did not depend on geometric figures. This book was most likely titled "The difficulties of arithmetic (Moškelāt al-hesāb)", and is not extant.

Based on the context, some historians of mathematics such as D. J. Struik, believe that Omar Khayyam must have known the formula for the expansion of the binomial {(a+b)^n}, where n is a positive integer. The case of power 2 is explicitly stated in Euclid's elements and the case of at most power 3 had been established by Indian mathematicians. Khayyam was the mathematician who noticed the importance of a general binomial theorem.

The argument supporting the claim that Khayyam had a general binomial theorem is based on his ability to extract roots.

One of Khayyam's predecessors, Al-Karaji, had already discovered the triangular arrangement of the coefficients of binomial expansions that Europeans later came to know as Pascal's triangle; Khayyam popularized this triangular array in Iran, so that Pascal's triangle is known there as Omar Khayyam's triangle.

In 1074–5, Omar Khayyam was commissioned by Sultan Malik-Shah to build an observatory at Isfahan and reform the Persian calendar. There was a panel of eight scholars working under the direction of Khayyam to make large-scale astronomical observations and revise the astronomical tables. Recalibrating the calendar fixed the first day of the year at the exact moment of the passing of the Sun's center across vernal equinox. This marks the beginning of spring or Nowrūz, a day in which the Sun enters the first degree of Aries before noon. The resultant calendar was named in Malik-Shah's honor as the Jalālī calendar, and was inaugurated on 15 March 1079. The observatory itself was disused after the death of Malik-Shah in 1092.

The Jalālī calendar was a true solar calendar where the duration of each month is equal to the time of the passage of the Sun across the corresponding sign of the Zodiac. The calendar reform introduced a unique 33-year intercalation cycle. As indicated by the works of Khazini, Khayyam's team implemented an intercalation system based on quadrennial and quinquennial leap years. Therefore, the calendar consisted of 25 ordinary years that included 365 days, and 8 leap years that included 366 days. The calendar remained in use across Greater Iran from the 11th to the 20th centuries. In 1911 the Jalali calendar became the official national calendar of Qajar Iran. In 1925 this calendar was simplified and the names of the months were modernized, resulting in the modern Iranian calendar. The Jalali calendar is more accurate than the Gregorian calendar of 1582, with an error of one day accumulating over 5,000 years, compared to one day every 3,330 years in the Gregorian calendar. Jalali calendar is the most perfect calendar ever devised.

According to Farabi's enumeration of the sciences, ʻilm al-nujūm, (erroneously translated by the Europeans as astrology), from at least the middle of the tenth century, was already split into two parts, one dealing with astrology and the other with theoretical mathematical astronomy. Nizami Aruzi of Samarcand relates that "I did not observe that he (Omar Khayyam) had any belief in astrological predictions, nor have I seen or heard of any of the great scientists who had such belief." While working for Sultan Sanjar as an astrologer, Omar Khayyam was asked to predict the weather – a job that he apparently did not do well.

A short treatise of Omar Khayyam is concerned with music theory in which he discusses the connection between music and arithmetic; Khayyam provided a systematic classification of musical scales, and discussed the mathematical relationship among notes, minor, major and tetrachords.

The earliest allusion to Omar Khayyam's poetry is from the historian Imad ad-Din al-Isfahani, a younger contemporary of Khayyam, who explicitly identifies him as both a poet and a scientist (Kharidat al-qasr, 1174).

One of the earliest specimens of Omar Khayyam's Rubiyat is from Fakhr al-Din Razi. In his work Al-tanbih ‘ala ba‘d asrar al-maw‘dat fi'l-Qur'an (ca. 1160), he quotes one of his poems (corresponding to quatrain LXII of FitzGerald's first edition).

Daya in his writings (Mirsad al-‘Ibad, ca. 1230) quotes two quatrains, one of which is the same as the one already reported by Razi. An additional quatrain is quoted by the historian Juvayni (Tarikh-i Jahangushay, ca. 1226–1283).

In 1340 Jajarmi includes thirteen quatrains of Khayyam in his work containing an anthology of the works of famous Persian poets (Munis al-ahrār), two of which have hitherto been known from the older sources. A comparatively late manuscript is the Bodleian MS. Ouseley 140, written in Shiraz in 1460, which contains 158 quatrains on 47 folia. The manuscript belonged to William Ouseley (1767–1842) and was purchased by the Bodleian Library in 1844.

In addition to the Persian quatrains, there are twenty-five Arabic poems attributed to Khayyam which are attested by historians such as al-Isfahani, Shahrazuri (Nuzhat al-Arwah, ca. 1201–1211), Qifti (Tārikh al-hukamā, 1255), and Hamdallah Mustawfi (Tarikh-i guzida, 1339).

There are a number of other Persian scholars who occasionally wrote quatrains, including Avicenna, Ghazzali, and Tusi. It is possible that for Khayyam poetry was an amusement of his leisure hours: "these brief poems seem often to have been the work of scholars and scientists who composed them, perhaps, in moments of relaxation".

FitzGerald's Rubaiyat of Omar Khayyam contains loose translations of quatrains from the Bodleian manuscript. It enjoyed such success in the fin de siècle period that a bibliography compiled in 1929 listed more than 300 separate editions, and many more have been published since.

Khayyam considered himself intellectually to be a student of Avicenna. According to Al-Bayhaqi, he was reading the metaphysics in Avicenna's the Book of Healing before he died.

There are six philosophical papers believed to have been written by Khayyam.

- One of them, On existence (Fi'l-wujūd), was written originally in Persian and deals with the subject of existence and its relationship to universals.
- Another paper, titled The necessity of contradiction in the world, determinism and subsistence (Darurat al-tadād fi'l-'ālam wa'l-jabr wa'l-baqā'), is written in Arabic and deals with free will and determinism.
- On being and necessity (Risālah fī'l-kawn wa'l-taklīf),
- The Treatise on Transcendence in Existence (Al-Risālah al-ulā fi'l-wujūd),
- On the knowledge of the universal principles of existence (Risālah dar 'ilm kulliyāt-i wujūd), and

- Abridgement concerning natural phenomena (Mukhtasar fi'l-Tabi'iyyāt).

Omar Khayyam's exhortations to drink wine should not be taken literally, but should be regarded rather in the light of Sufi thought where rapturous intoxication by "wine" is to be understood as a metaphor for the enlightened state.

Al-Qifti (ca. 1172–1248) in his work The History of Learned Men reports that Omar's poems were only outwardly in the Sufi style, but were written with an anti-religious agenda. He also mentions that he was at one point indicted for impiety, but went on a pilgrimage to prove he was pious. The report has it that upon returning to his native city he practiced a strictly religious life, going morning and evening to the place of worship.

Seyyed Hossein Nasr argues that it is "reductive" to use a literal interpretation of his verses (many of which are of uncertain authenticity to begin with) to establish Omar Khayyam's philosophy. Instead, he adduces Khayyam's interpretive translation of Avicenna's treatise Discourse on Unity (Al-Khutbat al-Tawhīd), where he expresses orthodox views on Divine Unity in agreement with the author. The prose works believed to be Omar's are written in the Peripatetic style and are explicitly theistic, dealing with subjects such as the existence of God and theodicy.

FitzGerald's translation was a factor in rekindling interest in Khayyam as a poet even in his native Iran. Sadegh Hedayat in his

Songs of Khayyam (Taranehha-ye Khayyam, 1934) reintroduced Omar's poetic legacy to modern Iran. Under the Pahlavi dynasty, a new monument of white marble, designed by the architect Houshang Seyhoun, was erected over his tomb.

A statue by Abolhassan Sadighi was erected in Laleh Park, Tehran, in the 1960s, and a bust by the same sculptor was placed near Khayyam's mausoleum in Nishapur. In 2009, the state of Iran donated a pavilion to the United Nations Office in Vienna, inaugurated at Vienna International Center.

Biographical Summary

Omar Khayyam was born in 1048 in Nishapur, a leading metropolis in Khorasan during medieval times that reached its zenith of prosperity in the eleventh century under the Seljuq dynasty. His full name, as it appears in the Arabic sources, was Abu'l Fath Omar ibn Ibrahim al-Khayyam. In Persian texts he is usually simply called Omar Khayyam. Although open to doubt, it has often been assumed that his forebears followed the trade of tent-making, since Khayyam means tent-maker in Arabic. The historian Bayhaqi, who was personally acquainted with Omar, provides the full details of his horoscope: "he was Gemini, the sun and Mercury being in the ascendant[...]". This was used by modern scholars to speculate his date of birth as 18 May 1048.

His boyhood was spent in Nishapur. His gifts were recognized by his early tutors who sent him to study under Imam Muwaffaq Nishaburi, the greatest teacher of the Khorasan region who tutored the children of the highest nobility. Khayyam was also taught by the mathematician, Abu Hassan Bahmanyar bin Marzban. After studying science, philosophy, mathematics and astronomy at Nishapur, about the year 1068 he traveled to the province of Bukhara, where he frequented the renowned library of the Ark.

In about 1070 he moved to Samarkand, where he started to compose his famous treatise on algebra under the patronage of Abu Tahir Abd al-Rahman ibn ʿAlaq, the governor and chief judge of the city.

Omar Khayyam was kindly received by the Karakhanid ruler Shams al-Mulk Nasr, who according to Bayhaqi, would "show him the greatest honor, so much so that he would seat [Omar] beside him on his throne".

In 1073–4 peace was concluded with Sultan Malik-Shah I who had made incursions into Karakhanid dominions. Khayyam entered the service of Malik-Shah in 1074–5 when he was invited by the Grand Vizier Nizam al-Mulk to meet Malik-Shah in the city of Marv.

Khayyam was subsequently commissioned to set up an observatory in Isfahan and lead a group of scientists in carrying out precise astronomical observations aimed at the revision of the Persian calendar. The undertaking began probably in 1076 and ended in 1079 when Omar Khayyam and his colleagues concluded their measurements of the length of the year, reporting it to 14 significant figures with astounding accuracy (less than 1 microsecond!).

After the death of Malik-Shah and his vizier, Omar fell from favor at court, and he soon set out on his pilgrimage to Mecca. A possible ulterior motive for his pilgrimage reported by Al-Qifti, was a public demonstration of his faith with a view to allaying the allegations of unorthodoxy leveled at him by a hostile clergy. He was then invited by the new Sultan Sanjar to Marv, possibly to work as a court astronomer. He was later allowed to return to Nishapur owing to his declining health. Upon his return, he seems to have lived the life of a recluse.

Omar Khayyam died at the age of 83 in his hometown of Nishapur on 4 December 1131, and he is buried in what is now the Mausoleum of Omar Khayyam. One of his disciples Nizami Aruzi relates the story that sometime during 1112–3 Khayyam was in Balkh in the company of Al-Isfizari (one of the scientists who had collaborated with him on the Jalali calendar) when he made a prophecy that "my tomb shall be in a spot where the north wind may scatter flowers over it". Four years after his death, Aruzi located his tomb in a cemetery in a then large and well-known quarter of Nishapur on the road to Marv. As it had been foreseen by Khayyam, Aruzi found the tomb situated at the foot of a garden-wall over which pear trees and peach trees had thrust their heads and dropped their flowers such that his tombstone was hidden beneath them.

42. Abu al-Salt

Abū al-Ṣalt Umayya ibn ʿAbd al-ʿAzīz ibn Abī al-Ṣalt al-Dānī al-Andalusī

(Arabic: أبو الصلت),

(c. 1068—October 23, 1134),

was an Andalusian-Arab polymath who wrote about pharmacology, geometry, physics, and astronomy.

Scientific Contributions

Abu al-Salt wrote an encyclopedic work of many treatises on the scientific disciplines known as quadrivium. This work was probably known in Arabic as Kitāb al-kāfī fī al-ʿulūm.

His interests also included alchemy as well as the study of medicinal plants.

Abī al-Ṣalt's research works on astronomical instruments were read both in the Islamic world and Europe.

Abū al-Ṣalt was a medical scientist and served as visiting physician in Palermo, Sicily, and worked in the court of Roger I of Sicily as a visiting physician. He became well known in Europe through translations of his works made in the Iberian Peninsula and in southern France. He is also credited with introducing Andalusian music to Tunis, which later led to the development of the Tunisian ma'luf.

Following are some astronomy research works of Abu al-Salt:

- Risāla fī al-amal bi-l-astrulab ("On the construction and use of the astrolabe")
- A description of the three instruments known as the Andalusian equatoria.
- Ṣifat ʿamal ṣafīḥa jāmiʿa taqawwama bi-hā jamīʿ al-kawākib al-sabʿa ("Description of the construction and Use of a Single Plate with which the totality of the motions of the seven planets"), where the seven planets refer to Mercury, Venus, earth, Moon, Mars, Jupiter, and Saturn.
- Kitāb al-wajīz fī ʿilm al-hayʾa ("Brief treatise on cosmology")
- Ajwiba ʿan masāʾil suʾila ʿan-ha fa-ajāba or Ajwiba ʿan masāʾil fī al-kawn wa-ʾl-ḥabīʿa wa-ʾl-ḥisāb ("Solution to questions on cosmology, physics, and arithmetic").
- An introduction to astronomy.
- A Summary of Ptolemy's Almagest.

Biographical Summary

Abu al-Salt was born in 1068 AD, in Dénia, al-Andalus. After the death of his father while he was a child, he became a student of al-Waqqashi (1017—1095) of Toledo (a colleague of Abū Isḥāq Ibrāhīm al-Zarqālī). Upon completing his mathematical education in Seville, he set out with his family to Alexandria and then Cairo in 1096.

In Cairo, he entered the service of the Fatimid ruler Abū Tamīm Ma'add al-Mustanṣir bi-llāh and the Vizier Al-Afdal Shahanshah. His service continued until 1108, when, according to Ibn Abī

Uṣaybiʿa, his attempt to retrieve a very large Felucca laden with copper, that had capsized in the Nile, ended in failure. Abu al-Salt had built a mechanical tool to retrieve the Felucca, and was close to success when the machine's silk ropes fractured. The Vizier Al-Afdal ordered Abu al-Salt's arrest, and he was imprisoned for more than three years; he was released in 1112 AD.

Abu al-Salt then left Egypt for Kairouan in Tunisia, where he entered the service of the Zirids in Ifriqiya.

He also occasionally worked as a visiting physician in Palermo, Sicily. He was in collaboration with the Palermo poet Abū l-Ḍawʾ.

He died in 1134 AD in Béjaïa, Algeria.

43. Ibn Bajja

Abū Bakr Muḥammad ibn Yaḥyà ibn aṣ-Ṣā'igh at-Tūjībī ibn Bājja (Arabic: أبو بكر محمد بن يحيى بن الصائغ التجيبي بن باجة),

(c. 1085 – 1138), is also known by his Latinised name Avempace. He was an Andalusian polymath, whose writings include works in astronomy, physics, medicine, botany, philosophy, music and poetry.

Scientific Contributions

Ibn Bājja was the author of the Kitāb an-Nabāt ("The Book of Plants"), a book on botany.

He defined the sex of plants.

In Botany, Ibn Bājja's work titled Kitab al-nabat (The Book of Plants) discusses the morphology of various plants and classifies them based on their similarities. He also writes about the reproductive nature of plants and their supposed genders based on his observations of palm and fig trees.

Kitab al-nabat has most recently been translated into Spanish.

Ibn Bājja's book Kitāb al-Tajribatayn 'alā Adwiyah Ibn Wāfid (Book of Experiments on Drugs of Ibn Wafid) classifies plants from a pharmacological perspective. It uses the work of Ibn al-Wafid, a physician.

In Physics, Ibn Bajja stated that for every force there is always a reaction force.

While he did not specify is that these forces be equal and opposite. It is an early version of the so-called Newton's Third Law of Motion. Newton's third law of motion states that for every action there is an equal and opposite reaction.

Ibn Bājja was a critic of Ptolemy and he worked on creating a new theory of velocity to replace the one theorized by Aristotle.

Two future philosophers supported the theories Ibn Bājja created, known as the Ibn Bājjaan dynamics. These philosophers that supported this dynamics were Thomas Aquinas, a Catholic priest, and John Duns Scotus.

Galileo Gallilei adopted Ibn Bājja's formula.

Though many of his works have not survived, his theories in astronomy and physics were preserved by Maimonides and Averroes, and influenced later astronomers and physicists in the Islamic civilization and Renaissance Europe, including Galileo Galilei.

In Astronomy, he asserts:

"The Milky Way is the light of many stars which almost touch one another. Their light forms a 'continuous image' (khayâl muttasil) on the surface of the body which is like a "tent" (takhawwum) under the fierily element and over the air which it covers. Ibn Bājja defines the continuous image as the result of refraction (inʿikâs) and supports its explanation with an observation of a conjunction of two planets, Jupiter and Mars which took place in 1106-7. He watched the conjunction and “saw them having an elongate figure” although their figure is circular."

Ibn Bājja also reported observing "two planets as black spots on the face of the Sun." In the 13th century, the Maragha astronomer Qotb al-Din Shirazi identified this observation as the transit of Venus and Mercury. However, Ibn Bājja cannot have observed a Venus transit, as there were no Venus transits in his lifetime.

Ibn Bājja worked under the mathematician Ibn al-Sayyid, and he later wrote a commentary to Ibn al-Sayyid's work on geometry and Euclid's Elements.

Ibn Bājja viewed astronomy as part of mathematics. Ibn Bājja's model of the cosmos consists of concentric circles, but no epicycles.

In Philosophy, Ibn Bājja’s theories were influential, though most of his writings and books were not completed (or well-organized) due to his early death.

He had a vast knowledge of medicine, mathematics, and astronomy. His main contribution to Islamic philosophy was his idea on soul phenomenology, which remained unfinished.

Ibn Bajja wrote the first commentaries on Aristotle. While his work on projectile motion was never translated from Arabic to Latin, his views became well known around the Western world and to Western philosophers, astronomers, and scientists of many disciplines. His works impacted contemporary thought, and later influenced Galileo and his work.

Ibn Bājja's theories on projectile motion are found in the text known as "Text 71"

Ibn Bājja was, in his time, not only a prominent figure of philosophy but also of music and poetry. His diwan (collection of poetry) was rediscovered in 1951.

Recently, the web page Webislam, created by Spanish converts to Islam, reported that the score of the Nuba al-Istihlál of Ibn Bājja (11th century), arranged by Omar Metiou and Eduardo Paniagua, is very similar to Marcha Granadera (18th century), which is now the official anthem of Spain. That makes it the world's oldest song (about a thousand years old) used for the official anthem of a country.

In 2009, a crater 199 km (62 mi) from the South Pole of the Moon was designated the Ibn Bajja crater by the International Astronomical Union (IAU) in his honor.

Some of Ibn Bajja's other works are listed below.

Upon his unplanned trip to Egypt, Ibn Bājja wrote Risālat al-wadāʿ (Letter of Bidding Farewell)

Risālat al-ittiṣāl al-ʿaql bi al-insān (Letter on the Union of the Intellect with Human Beings) that were dedicated to Ibn al-Imām.

Tadbīr al-mutawaḥḥid (Management of the Solitary).

Kitāb al-nafs (Book on the Soul).

Risāla fī l-Ghāya al-insāniyya (Treatise on the Objective of Human Beings).

Some of his famous accomplishments were written near the end of his life.

Biographical Summary

Ibn Bājja was born in Zaragoza, in what is today Aragon, Spain, around 1085AD. He died in Fez, Morocco, in 1138.

Ibn Bājja also had a talent for singing and composition in music. In the beginning of his career, he wrote the manuscript Risālah fī l-alḥān (Tract on melodies) and incorporated his commentary on al-Fārābī's treatise on music. He determined the correlations between different melodies and temperament. According to biographer al-Maqqarī, Ibn Bājja's passion for music was due to poetry and had "the virtue of dispelling the sadness and pain of the hearts." He included his scientific knowledge and wit in many poems. Ibn Bājja joined in poetic competitions with the poet al-Tutili.

Rulers of Zaragoza shifted constantly throughout Ibn Bājja's young life. Ali Ibn Yusuf Ibn Tashfin was the Almoravid Sultan. In 1114, Almoravid Sultan appointed Abu Bakr 'Ali ibn Ibrahim as-Sahrawi (also known as Ibn Tifilwit) as the governor of Zaragoza.

Music and poetry brought a close relationship between Ibn Bājja and Ibn Tifilwit. This is verified in writings by both Ibn al-Khatib and Ibn Khaqan. Ibn Bājja enjoyed music and poetry with the governor. Ibn Bājja was appointed a vizier to Ibn Tifilwit.

There was a diplomatic mission to meet the overthrown Imad ad-Dawla Ibn Hud king in his castle. Perhaps in this connection, Ibn Bājja was placed in jail for some months.

Ibn Tifilwit was killed during a campaign against the Christians in 1116, ending his short reign. That inspired Ibn Bājja to compose mournful elegies in his honor.

After the fall of Zaragoza in 1118, Ibn Bājja looked for shelter under Abu Ishaq Ibrahim ibn Yûsuf Ibn Tashfin who was the governor of Xàtiba and a brother of Almoravid Sultan. Ibn Bājja served as the vizier of Yusuf Ibn Tashfin for some twenty years. Throughout these decades, it is clear that Ibn Bājja was not as agreeable with those close to the ruler, Ibn Tashfin, as he was during the previous reign of Ibn Tifilwit.

Writings by Ahmad al-Maqqari give us insight into the disagreements between Ibn Bājja and the father (Abd al-Malik) of a famous physician respected by Ibn Tashfin. A poetry anthology, Qala'id al-iqya (Necklace of Rubies), was created by a courtier of Ibn Tashfin, namely Abu Nasr al-Fath Ibn Muhammad Ibn Khaqan. In this book Ibn Bājja was condescendingly placed in the last place. Ibn Bājja was imprisoned twice Under Ibn Tashfin, the Sultan of the Almoravid empire.

Despite being unwelcome, Ibn Bājja remained with the Almoravid empire for the rest of his life until his death in 1138 AD in Fez.

Al-Maqqari details in his writing that a physician, Abu l-'Ala' Ibn Zuhr, was an enemy of Ibn Bājja whose servant, Ibn Maʿyub, was suspected of poisoning Ibn Bājja. However, this allegation was not proven.

44. Muhammad al-Baghdadi

Abu Bakr Muhammad Ibn ‘Abd al-Bāqī al-Baghdadi al-Ansārī al-Kaabī

(Arabic: محمد بن عبد الباقي البغدادي),

(1050-1141),

was a jurist and a mathematician.

Scientific Contributions

Muhammad Al-Baghdadi authored a commentary on the tenth book of Euclid's Elements, which was translated by Gerard of Cremona as Liber judei super decimum Euclidis. The work was popular in Europe with several Latin manuscripts still extant.

Other works include:

- *Jadawil al-Jayb al-Mahlul al-Daqiqa* (detailed tables of sines for each minute),
- Risala fi Taqrib Usul al-Hisab fi' al-Jabr wa-‘l-Muqabala (Treatise on approximation of principles of Algebra)
- Kitab al-Tabaqat fi Sharh al-Misaha (book on measurements)

Biographical Summary

Muhammad Al-Baghdadi was born in 1050 AD and he died in 1141 AD.

45. Ibn al-'Awwam

Abu Zakariya Ibn al-Awwam

(Arabic: أبو زكريا بن العوام), also known as Abū l-Khayr al-Ishbīlī, was an agriculturist from Seville, Andalusia, in the later 12th century.

Scientific Contributions

Ibn al-Awwam wrote a lengthy handbook on agriculture entitled Kitāb al-Filāḥa (Book on Agriculture), which is the most comprehensive treatment of the subject in the world, in any language.

It was published in Spanish and French translations in the 19th century. The edition in French is 1350 pages.

Kitāb al-Filāḥa Al-'Awwam cites information from 112 different prior authors, with 1900 total citations. He had read the Book of Nabataean Agriculture by Ibn Wahshiyya; and other Andalusian authors like Ibn Bassal, Abu al-Khayr al-Ishbili and Ibn Hajjaj. They were in the later 11th century in southern Spain; and copies of their works have survived only partly and incompletely). His most influential direct sources were the Andalusian Arabic sources.

Ibn al-Awwam's book is divided into thirty-four chapters.

- The first four chapters in the book deal successively with different types of soils, fertilizers, irrigation, and planning a garden layout.

- Then there are five chapters on growing fruit trees, including grafting, pruning, growing from cuttings, etc., and dozens of different fruit trees are treated individually.
- Later chapters deal with ploughing, the choice of seeds, the seasons and their tasks, grain farming, leguminous plants, small allotments, aromatic plants and industrial plants. Again, many plants are treated individually on how to cultivate them.
- One chapter is devoted to methods of preserving and storing foods after harvest, a topic which also comes up intermittently elsewhere. The symptoms of many diseases of trees and vines are indicated, as are methods of cure.
- The above together make up 30 chapters of the book, together they deal with crops. The last four chapters deal with livestock including discussion of the diseases and injuries to horses and cattle and their cures.

Ibn al-Awwam's book is an encyclopedia on agriculture. It is an encyclopedia that is guided and informed by Ibn al-Awwam's own research on agriculture.

Ibn al-Awwam was a pioneering agriculturalist and a botanist.

Kitāb al-Filāḥa was published in 1802 with the Arabic text placed alongside a translation into Spanish, and in 1864 it was published in French. These publications are freely available online.

Biographical Summary

His full name was Abu Zakariya Yahya ibn Muhammad ibn Ahmad ibn Al-'Awwam Al-Ishbili (Arabic: أبو زكريا يحيى بن محمد بن أحمد بن العوام). He was born in 12th century in Seville. His dates of birth and death are not known.

Nearly everything that is known about his biography is gleaned from his book. It appears that he may have been a landowner whose interests lay in agricultural matters. It is clear that he did much hands-on research and experimentation with a wide range of crops, animal husbandry, and horticulture. It is also clear that he was well-read.

46. Jabir ibn Aflah

Abū Muḥammad Jābir ibn Aflaḥ

(Arabic: أبو محمد جابر بن أفلح),

(1100–1150),

was an astronomer and mathematician from Seville, in 12th century al-Andalus.

Scientific Contributions

Jābir ibn Aflaḥ's work Iṣlāḥ al-Majisṭi (Correction of the Almagest) was an influential work in astronomy.

He invented an observational instrument known as the torquetum, a research device to transform between spherical coordinate systems.

One substantial change Jābir made is that he placed the orbits of Venus and Mercury outside, rather than between the Moon and the Sun.

Iṣlāḥ al-Majisṭi is a first criticism of Almagest. He particularly criticized the mathematical basis of the work. For example, he based the calculations on spherical trigonometry.

The work was translated from the Arabic into both Hebrew and Latin, the latter by Gerard of Cremona, who Latinized his name as "Geber". Through that channel, the work of Jābir ibn Aflaḥ had a wide influence on later European mathematicians and astronomers and helped to promote trigonometry in Europe.

The 16th century writing by Gerolamo Cardano showed that much of the material on spherical trigonometry "On Triangles" by Regiomontanus (c.1463) *was taken directly and without credit from Jābir's work.* Such has happened with numerous other works also.

Biographical Summary

Jābir ibn Aflaḥ was born in 1100 AD in Andalusia and he died in 1150.

47. Abu al-Fadl Al-Dimashqi

Abū al-Faḍl Jaʿfar ibn ʿAlī al-Dimashqī

(Arabic: أبو الفضل جعفر بن علي الدمشقي),

(12th-century),

was a prosperous merchant and an economic theorist from Damascus.

Scientific Contributions

Al-Dimashqi is world's first economic theorist.

He is best known for being the author of:

Kitab al-Isharah ila Mahasin at-Tijarah wa Marifat Jayyid al-A'rad wa Kadiiha wa Ghush-ush al-Mudallisin fiha

(A Guide to the Merits of Commerce and to Recognition of Both Fine and Defective Merchandise and the Swindles of Those Who Deal Dishonestly).

Al-Dimashqi's economic research work praises trade as an economic activity and demonstrates a thorough understanding of the roles of supply and demand and the uncertainty associated with them. He also formulated a theory of pricing in trade. He focuses on the quality of merchandise and the very first analysis in economics on the subject of "quality control".

Al-Dimashqi's theory of economics introduces the basic concepts of costs of production, demand for the product, supply of the product, scarcity of the product, product quality control, and the methods of market pricing including the speculative trade practices.

Al-Dimashqi explicitly introduces the role of division of labor (specialization) for economic development. He also recognizes a role for the state in the economic development.

Following is an excerpt from Al-Dimashqi's economic theory.

"No individual can, because of the shortness of his life span, burden himself with all industries. If he does, he may not be able to master the skills of all of them from the first to the last. Industries are all interdependent. Construction needs the carpenter and the carpenter needs the ironsmith and the ironsmith needs the miner, and all these industries need premises. People are, therefore, necessitated by force of circumstances to be clustered in cities to help each other in fulfilling their mutual needs".

Biographical Summary

Al-Dimashqi was a 12th century merchant who lived in Damascus. Almost nothing else is known about al-Dimashqi's life.

48. Muhammad Al-Idrisi

Abu Abdullah Muhammad al-Idrisi al-Qurtubi al-Hasani as-Sabti, (or simply al-Idrisi)

(Arabic: أبو عبد الله محمد الإدريسي القرطبي الحسني السبتي),

(1100 – 1165),

was a geographer, cartographer and Egyptologist.

Scientific Contributions

Al-Idrisi authored his geographical treatise titled: Kitāb Nuzhat al-mushtāq fī dhikr al-amṣār wa-al-aqṭār wa-al-buldān wa-al-juzur wa-al-madā' in wa-al-āfāq. The title has been translated as The book of pleasant journcys into faraway lands. It has been preserved in nine manuscripts, seven of which contain maps.

In the introduction, al-Idrisi mentions two sources for geographical coordinates: Claudius Ptolemy and Ishaq ibn al-Hasan al-Zayyat. Al-Idrisi cross-checked oral reports from different informers to see if these geographical coordinates were consistent.

An abridged version of the Arabic text was published in Rome in 1592 with title: De geographia universali. The first translation from the original Arabic was into Latin. The Maronite's Gabriel Sionita and Joannes Hesronita translated an abridged version of the text which was published in Paris in 1619 with the title: Geographia nubiensis.

Not until the middle of the 19th century was a complete translation of the Arabic text published. This was a translation into French by Pierre Amédée Jaubert. More recently sections of the text have been translated for particular regions. Beginning in 1970 a critical edition of the complete Arabic text was published.

Al-Idrisi's Nuzhat al-Mushtaq states the following on the Atlantic Ocean:

The Commander of the Muslims Ali ibn Yusuf ibn Tashfin sent his admiral Ahmad ibn Umar, better known under the name of Raqsh al-Auzz to attack a certain island in the Atlantic, but he died before doing that. [...] Beyond this ocean of fogs it is not known what exists there. Nobody has the sure knowledge of it, because it is very difficult to traverse it. Its atmosphere is foggy, its waves are very strong, its dangers are perilous, its beasts are terrible, and its winds are full of tempests.

There are many islands, some of which are inhabited, others are submerged. No navigator traverses them but bypasses them remaining near their coast. [...] And it was from the town of Lisbon that the adventurers set out known under the name of Mughamarin [Adventurers], penetrated the ocean of fogs and wanted to know what it contained and where it ended. [...] After sailing for twelve more days they perceived an island that seemed to be inhabited, and there were cultivated fields. They sailed that way to see what it contained. But soon barques encircled them and made them prisoners, and transported them to a miserable hamlet situated on the coast. There they landed. The navigators saw their people with red skin; there was not much hair on their body, the hair of their head was straight, and they were of high stature. Their women were of an extraordinary beauty.

(translation by Professor Muhammad Hamidullah)

This description matches "Red Indians". It is plausible to argue that this Muslim expedition made it to Red Indian territory more than three centuries before Columbus did the same. There is, however, one important difference: Muslims went there as expeditors and explorers, not to claim the lands.

As a cartographer, Al-Idrisi prepared a detailed world map: Tabula. In the map Al-Idrisi describes Irlandah-al-Kabirah (Great Ireland) as one day's sailing distance from the extremity of Iceland to that of Great Ireland.

Al-Idrisi describes that Chinese junks (special kind of ships) carried leather, swords, iron and silk. He mentions the glassware of the city of Hangzhou and labels Quanzhou's silk as the best. In his records of Chinese trade, al-Idrisi also wrote about the Silla Dynasty (one of Korea's historical Dynasties, and a major trade partner to China at the time).

Al-Idrisi also authored a Medical dictionary. Al-Idrisi organized a list of drugs and plants and their curative effects. These were used by physicians, apothecaries and merchants.

The dictionary is unique, as it includes the names of drugs in as many as 12 languages (among which are Spanish, Berber, Latin, Greek and Sanskrit). At the end of the section on medicinal herbs which are alphabetically organized, he gives an index of their entries.

Biographical Summary

Al-Idrisi was born in 1099 AD in the city of Ceuta, at the time ruled by the Almoravids. He was from the large Hammudid dynasty of North Africa and Al-Andalus, descendants from the Idrisi's of Morocco.

Al-Idrisi visited Anatolia when he was barely 16. He spent much of his early life travelling through North Africa and Al-Andalus (Muslim Spain) and acquired detailed information on both regions. He also traveled to many parts of Europe including Portugal, the Pyrenees, the French Atlantic coast, Hungary, and Jórvík (now known as York).

Al-Idrisi studied in Córdoba, a great center of learning.

Al-Idrisi died in 1166 AD.

49. Ibn al-Kammad

Abu Jafar Ahmad ibn Yusuf ibn al-Kammad

(Arabic: أحمد بن يوسف ابن الكماد),

(died 1195),

was an astronomer born in Seville, Al-Andalus.

Scientific Contributions

Ibn al-Kammad was educated in Cordoba by the students of Al-Zarqali. His works include al Kawr ala al dawr, al Amad ala al abad, and al Muqtabas. His Zij was very well received not only on the Iberian peninsula but across North Africa, specifically Tunisia.

The astronomer Ibn Ishaq al-Tunisi wrote commentaries on his works.

Kitāb Mafātīḥ alasrār of Ibn al-Kammad states that the duration of pregnancy can depend on the star configurations, though it was criticized by Ibn al-Haim al-Ishbili in the latter's al Zīj al kāmil (c.1205).

Biographical Summary

Ibn al-Kammad was born in Seville, Al-Andalusia. He died in 1195 AD.

50. Nur ad-Din al-Bitruji

Nur ad-Din al-Bitruji

(Arabic: أبو إسحاق البطروجي), (also spelled Nur al-Din Ibn Ishaq al-Betrugi and Abu Ishâk ibn al-Bitrogi) (known in the West by the Latinized name of Alpetragius),

(died c. 1204),

was an Iberian-Arab astronomer and a Qadi in al-Andalus.

Scientific Contributions

Al-Biṭrūjī was the first astronomer to present a non-Ptolemaic astronomical system as an alternative to Ptolemy's models, with the planets borne by geocentric spheres. Another original aspect of his system was that *he proposed a physical cause of celestial motions.* His alternative system spread through most of Europe during the 13th century.

Al-Bitruji wrote Kitāb al-Hayʾah (The book of theoretical astronomy/cosmology, Arabic, كتاب الهيئة), which presented criticism of Ptolomy's Almagest from a physical point of view. It was well known in Europe between the 13th and the 16th centuries, as superseding Ptolemy's Almagest.

This work was translated into Latin by Michael Scot in 1217 as De motibus celorum (first printed in Vienna in 1531). A Hebrew translation by Moses ibn Tibbon was done in 1259.

There is also an anonymous treatise on tides (Escorial MS 1636, dated 1192) which contains material borrowed from al-Bitruji.

Al-Bitruji proposed a theory on planetary motion in which he avoided both epicycles and eccentrics, and accounted for the phenomena peculiar to the wandering stars, by compounding rotations of homocentric spheres.

This was a modification of the system of planetary motion proposed by his predecessors, Ibn Bajjah (Avempace) and Ibn Tufail (Abubacer).

The numerical predictions of the planetary positions in his theory were less accurate than those of the Ptolemaic model; however, his theory simplified the calculations by eliminating the need to map Ptolemy's epicycles onto Aristotle's concentric spheres.

His theory is an update and reformulation of that of Eudoxus of Cnidus, combined with the motion of fixed stars developed by al-Zarqālī: however, Al-Bitruji had no access or knowledge of Eudoxus' work.

A new original aspect of al-Biṭrūjī's system is his proposal of a physical cause of celestial motions. He combines the idea of "impetus" (first proposed by John Philoponus) and the concept of "shawq" (first proposed by Abū al-Barakāt al-Baghdādī) to *explain how energy is transferred from a first mover placed in the 9th sphere to other spheres*, explaining the other spheres' variable speeds and different motions.

He discarded the Aristotelian idea that there is a specific kind of dynamics for each world, *applying instead the same dynamics* to the sublunar and the celestial worlds.

This theory is a precursor to the work of Newton on gravity and the calculation of the planetary orbits as the conic sections using the same law throughout.

His theory spread through most of Europe during the 13th century, with debates and refutations of his ideas continued up to the 16th century. Copernicus cited his system in the De revolutionibus while discussing theories of the order of the inferior planets.

Biographical Summary

Almost nothing about Nur ad-Din al-Bitruji's life is known, except that his name probably derives from Los Pedroches (al-Biṭrawsh), a region near Cordoba. He was a disciple of Ibn Tufail (Abubacer) and was a contemporary of Averroes.

Nur ad-Din al-Bitruji died in 1204 AD.

51. Ismail Al-Jazari

Badīʿ az-Zaman Abu l-ʿIzz ibn Ismāʿīl ibn ar-Razāz al-Jazarī

(Arabic:بديع الزمان أَبُ اَلْعِزِ إبْنُ إسْماعِيلِ إبْنُ الرّزاز الجزري),

(1136–1206),

was a polymath: a scholar, inventor, mechanical engineer, artisan, artist and mathematician.

Scientific Contributions

Al-Jazari is known for writing in 1206 "The Book of Knowledge of Ingenious Mechanical Devices" (Arabic: كتاب في معرفة الحيل الهندسية). In this book Al-Jazari described 50 mechanical devices, along with instructions on how to construct them.

He invented the elephant clock.

Al-Jazari is the "father of robotics" and modern-day mechanical engineering.

Al-Jazari invented the fundamental principles and components of mechanical engineering. Following are some of these principles and components.

Camshaft.

Al-Jazari introduced camshaft into mechanical engineering in 1206. He used this invention to enable automation of water clocks (such as the candle clock) and water-raising machines.

The camshaft appeared in European mechanisms three centuries later, from the 14th century.

Crankshaft and crank-slider mechanism.

In 1206, al-Jazari invented a crankshaft, which he incorporated with a crank-connecting rod mechanism in his twin-cylinder pump. Like the modern crankshaft, al-Jazari's mechanism consisted of a wheel setting several crankpins into motion, with the wheel's motion being circular and the pins moving back-and-forth in a straight line. The crankshaft described by al-Jazari transforms continuous rotary motion into a linear reciprocating motion, and is central to modern machinery such as the steam engine, internal combustion engine as well as the automatic controls.

He used the crankshaft with a connecting rod in two of his water-raising machines: the crank-driven saqiya chain pump and the double-action reciprocating piston suction pump.

His water pump also employed the first known crank-slider mechanism.

Design and construction methods.

We see for the first time in al-Jazari's work several concepts important for both design and construction: the lamination of timber to minimize warping, the static balancing of wheels, the use of wooden templates (a kind of pattern), the use of paper models to establish designs, the calibration of orifices, the grinding of the seats and plugs of valves together with emery powder to obtain a watertight fit, and the casting of metals in closed mold boxes with sand.

Escapement mechanism in a rotating wheel.

Al-Jazari invented a method for controlling the speed of rotation of a wheel using an escapement mechanism.

Mechanical controls.

Al-Jazari described several early mechanical controls, including "a large metal door, a combination lock and a lock with four bolts".

Segmental gear.

A segmental gear is "a piece for receiving or communicating reciprocating motion from or to a cogwheel, consisting of a sector of a circular gear, or ring, having cogs on the periphery, or face."

Segmental gears first clearly appear in al-Jazari, in the West they emerge in Giovanni de Dondi's astronomical clock finished in 1364, and only with the great Sienese engineer Francesco di Giorgio (1501) did they enter the general vocabulary of European machine design.

Al-Jazari is the father of modern mechanical engineering. He clearly discusses details about the machines: mechanisms, components, ideas, methods, and design features. He demonstrated in detail how these are deployed to generate the machines using principles of mechanical engineering.

Following are some of the machines Al-Jazari invented based on these principles and fundamental components.

Water-raising machines

Al-Jazari invented five machines for raising water, as well as watermills and water wheels with cams on their axle used to operate automata. He described them in 1206. It was in these water-raising machines that he introduced his most important ideas and components.

Saqiya chain pumps.

The first known use of a crankshaft in a chain pump was in one of al-Jazari's saqiya machines. The concept of minimizing intermittent working is also first implied in one of al-Jazari's saqiya chain pumps, which was for the purpose of maximising the efficiency of the saqiya chain pump. Al-Jazari also constructed a water-raising saqiya chain pump which was run by hydropower rather than manual labor. Saqiya machines like the ones he described have been supplying water in Damascus since the 13th century up until modern times, and were in everyday use throughout the Islamic world.

Double-action suction pump with valves and reciprocating piston motion.

Al-Jazari described his version of suction pipes, suction pump, double-action pump, and made early uses of valves and a crankshaft-connecting rod mechanism. He used them to develop a twin-cylinder reciprocating piston suction pump. This pump is driven by a water wheel, which drives, through a system of gears, an oscillating slot-rod to which the rods of two pistons are attached. The pistons work in horizontally opposed cylinders, each provided with valve-operated suction and delivery pipes. The delivery pipes are joined above the center of the machine to form a single outlet into the irrigation system. This water-raising machine had a direct significance for the development of modern engineering. This pump is remarkable for three reasons:

- The first known use of a true suction pipe (which sucks fluids into a partial vacuum) in a pump.
- The first application of the double-acting principle.
- The conversion of rotary to reciprocating motion via the crank-connecting rod mechanism.

Al-Jazari's suction piston pump could lift 13.6 meters of water with the help of delivery pipes.

Water supply system.

Al-Jazari developed the earliest water supply system to be driven by gears and hydropower, which was built in 13th century Damascus to supply water to its mosques and Bimaristan (hospitals). The system had water from a lake turn a scoop-wheel and a system of gears which transported jars of water up to a water channel that led to mosques and hospitals in the city.

Automation

Al-Jazari built automated moving peacocks driven by hydropower. He also invented the earliest known automatic gates. These were driven by hydropower, and operated automatic doors as part of one of his elaborate water clocks. He invented water wheels with cams on their axle to operate automation.

The Italian Renaissance inventor Leonardo da Vinci used this automation technology invented by Al-Jazari.

Robotics

Muslim scientists displayed an interest in creating human-like machines (robotics) for serving humanity.

Drink-serving waitress

One of al-Jazari's humanoid automata was a waitress that could serve water, tea, or drinks. The drink was stored in a tank with a reservoir from where the drink drips into a bucket and, after seven minutes, into a cup, after which the waitress appears out of an automatic door to serve the drink.

Hand-washing automation with flush mechanism

Al-Jazari invented a hand washing automaton incorporating a flush mechanism now used in modern flush toilets. This device is another example of humanoid automata. It consisted of a human figure, made from jointed copper, holding a pitcher resembling a peacock in its right hand. The pitcher is made from brass and holds within it a chamber, divided into two parts by a metal plate. This mechanism aided the pouring of the water from the spout so that it was smooth and would not splutter. The reservoir in which the water is held is situated within the right-hand side of the human figure. An axle is fitted into the right elbow of the human figure so as to allow the liquid to pour from the reservoir through the spout of the pitcher. The left arm of the figure had a fixed weight which would raise and lower the arm which would hold a towel, comb and mirror.

This automaton was designed to aid the king whilst he performed his ritual ablutions. A servant of the king would carry the figure and place it next to a basin that could hold liquid. The servant then turned a knob on the back of the figure which opened a valve resulting in the pouring of water from the right hand of the figure into the basin. When the reservoir is nearly empty and most of the water has been poured a mechanism is prompted and the left hand of the figure, holding the towel, comb and mirror, is extended out in the direction of the king so that he can dry himself and tend to his beard.

Peacock fountain with automated servants

Water and its usages hold particular importance in Islam; both as being an integral part of the pre-prayer washing processes, wudu, and ghusl; and a key feature in Islamic gardens, four fountains featuring in the Garden of Eden referenced to in the Quran.

Additionally, with Mesopotamia being a naturally drought-ridden place, machines relating to water held a significant function; in both a divine and practical sense.

An entire section of The Book of Knowledge of Ingenious Mechanical Devices was devoted to fountain mechanisms, titled: 'On the construction in pools of fountains which change their shape, and of machines for the perpetual flute'.

Al-Jazari's "peacock fountain" was a more sophisticated hand washing device featuring humanoid automata as servants which offer soap and towels.

Pulling a plug on the peacock's tail releases water out of the beak; as the dirty water from the basin fills the hollow base, a float rises and actuates a linkage which makes a servant figure appear from behind a door under the peacock and offer soap. When more water is used, a second float at a higher-level trips and causes the appearance of a second servant figure, with a towel!

The basin of the "peacock fountain" formed the basin for performing wudu, and it would have been operated by a servant, who would have pulled the plug and positioned the peacock's beak; allowing the mechanism to release the water into the basin in front of the user.

Musical robot band

Al-Jazari's work described fountains and musical automata, in which the flow of water alternated from one large tank to another at hourly or half-hourly intervals. This operation was achieved through Al-Jazari's innovative use of hydraulic switching.

Al-Jazari created a musical automaton, which was a boat with four automatic musicians that floated on a lake to entertain guests at royal parties. It had a programmable drum machine with pegs (cams) that bump into little levers that operated the percussion. The drummer could be made to play different rhythms and different drum patterns if the pegs were moved around.

The water-clock of the drummers

The water-clock of the drummers, which differs from the Musical robot band in that it lacks a flute-playing doll and instead has two trumpeters, consists of seven wood-jointed male figures, including the aforementioned trumpeters as well as two dolls playing cymbals and the rest playing other percussive instruments. The mechanism in this specific automaton serves as a clock by producing a musical output once every hour, illustrating Al-Jazari's ability to create multi-faceted automata. The motion of the automaton is initiated at daybreak by another male doll, who stands at the edge of the frieze element of the design, moving across until he reaches a specific point at which a carved falcon leans forward dropping a ball from its beak onto a cymbal. All mechanical aspects of the automaton are then driven by water and a series of pistons and cables. Each hour water drains out of the main cistern to cause another bucket to tip over driving a water wheel that is connected to the musicians. The automaton is described to 'perform with a clamorous sound which is heard from afar' and could play several different tunes. Like many other automatons by Al-Jazari, this was created to entertain guests at the royal palace.

Al-Jazari constructed a variety of water clocks and candle clocks. These included a portable water-powered scribe clock, which was a meter high and half a meter wide, reconstructed successfully at the Science Museum in 1976.

Al-Jazari also invented monumental water-powered astronomical clocks which displayed moving models of the Sun, Moon, and stars.

Castle clocks

Al-Jazari's largest astronomical clock was the "castle clock", which was a complex device about 11 feet high, and had multiple functions besides timekeeping. It included a display of the zodiac and the solar and lunar orbits, and an innovative feature of the device was a pointer in the shape of the crescent moon which travelled across the top of a gateway, moved by a hidden cart, and caused automatic doors to open, each revealing a mannequin, every hour.

An innovative feature was the ability to reprogram the length of day and night in order to account for their changes throughout the year.

Another feature of the device was five automata musicians who automatically play music when moved by levers operated by a hidden camshaft attached to a water wheel. Other components of the castle clock included a main reservoir with a float, a float chamber and flow regulator, plate and valve trough, two pulleys, crescent disc displaying the zodiac, and two falcon automata dropping balls into vases.

Al-Jazari's castle clock is considered to be the *earliest programmable analog computer.*

Weight-driven water clocks

Al-Jazari invented water clocks that were driven by both water and weights. These included geared clocks and a portable water-powered scribe clock, which was a meter high and half a meter wide. The scribe with his pen was synonymous to the hour hand of a clock.

Al-Jazari's famous water-powered scribe clock was reconstructed successfully at the Science Museum, London in 1976.

Miniature paintings

Alongside Al-Jazari's accomplishments as an inventor and engineer, al-Jazari was also an accomplished artist. In The Book of Knowledge of Ingenious Mechanical Devices, he gave instructions of his inventions and illustrated them using miniature paintings, in a style of Islamic art.

Biographical Summary

Al-Jazari was born in the area of Upper Mesopotamia in 1136. He was born in Jazirat ibn Umar, or Al-Jazira for short, which was used to denote Upper Mesopotamia. He got the name Al-Jazari from his place of birth, Al-Jazira. The only biographical information known about him is contained in his Book of Knowledge of Ingenious Mechanical Devices.

Like his father before him, he served as chief engineer at the Artuklu Palace, the residence of the Mardin branch of the Artuqids which ruled across eastern Anatolia as vassals of the Zengid dynasty of Mosul and later of Salahuddin Ayyubid.

Al-Jazari's Book of Knowledge of Ingenious Mechanical Devices appears in a large number of manuscript copies, and as he explains repeatedly, he only describes devices he has built himself.

Al-Jazari goes on to describe the improvements he made to the work of his predecessors, and describes a number of devices, techniques and components that are original innovations which do not appear in the works by his predecessors.

Al-Jazari died in 1206 AD.

52. Sharaf al-Din al-Tusi

Sharaf al-Dīn al-Muẓaffar ibn Muḥammad ibn al-Muẓaffar al-Ṭūsī (Persian: شرف‌الدین مظفر بن محمد بن مظفر توسی),

(c. 1135 – c. 1213),

was an Iranian mathematician and astronomer

Scientific Contributions

Sharaf al-Dīn al-Tusi proposed the idea of a function in algebra. However, the approach is commonly attributed to Gottfried Leibniz.

Al-Tusi invented a method to numerically approximate the root of a cubic equation. He also developed a novel method for determining the conditions under which certain types of cubic equations would have two, one, or no solutions. This method was centuries later called the "Ruffini-Horner method".

The equations in question can be written, f(x) = c, where f(x) is a cubic polynomial in which the coefficient of the cubic term x3 is −1, and c is positive. The Muslim mathematicians of the time divided the potentially solvable cases of these equations into five different types, determined by the signs of the other coefficients in f(x). For each of these five types, al-Tusi wrote down an expression m for the point where the function f(x) attained its maximum, and gave a geometric proof that f(x) < f(m) for any positive x different from m.

He then concluded that:

the equation would have two solutions if $c < f(m)$,

one solution if $c = f(m)$,

or none if $c > f(m)$.

The quantities $D = f(m) - c$ which can be obtained from al-Tusi's conditions for the number of roots of cubic equations by subtracting one side of these conditions from the other is today called the discriminant of the cubic polynomials obtained by subtracting one side of the corresponding cubic equations from the other.

The discovery of these conditions by Sharaf al-Dīn al-Tusi demonstrated an understanding of the importance of the discriminant for investigating the solutions of cubic equations.

Sharaf al-Dīn al-Tusi analyzed the equation $x3 + d = b \cdot x2$ in the form $x2 \cdot (b - x) = d$, stating that the left-hand side must at least equal the value of d for the equation to have a solution. He then determined the maximum value of this expression. A value less than d means no positive solution; a value equal to d corresponds to one solution, while a value greater than d corresponds to two solutions. Sharaf al-Dīn al-Tusi's analysis of this equation was a notable development in mathematics.

Sharaf al-Din al-Tusi's "Treatise on equations" inaugurates the beginning of algebraic geometry.

Sharaf al-Din al-Tusi invented a linear astrolabe, sometimes called the "staff of Tusi". While it was easier to construct and was known in al-Andalus, it did not gain popularity.

Biographical Summary

Sharaf al-Din al-Tusi was probably born in 1135 AD in Tus, Iran. Little is known about his life, except what is found in the biographies of other scientists and that many mathematicians today can trace their lineage back to him.

Around 1165, he moved to Damascus and taught mathematics there. He then lived in Aleppo for three years, before moving to Mosul, where he met his most famous disciple Kamal al-Din ibn Yunus (1156-1242). This Kamal al-Din would later become the teacher of another famous mathematician from Tus, Nasir al-Din al-Tusi.
According to Ibn Abi Usaibi'a, Sharaf al-Din was "outstanding in geometry and the mathematical sciences, having no equal in his time". Sharaf al-Din al-Tusi died in 1213 AD.

53. Fibonacci

Fibonacci

also known as Leonardo Bonacci, Leonardo of Pisa, or Leonardo Bigollo Pisano ('Leonardo the Traveller from Pisa'),

(c. 1170 – c. 1240–50),

was an Italian mathematician from the Republic of Pisa, considered to be "the most talented Western mathematician of the Middle Ages".

Note: The name he is commonly called, Fibonacci, was made up in 1838 by the Franco-Italian historian Guillaume Libri and is short for "filius Bonacci" ('son of Bonacci'). However, even earlier in 1506 a notary of the Holy Roman Empire, Perizolo, mentions Leonardo as "Lionardo Fibonacci".

Scientific Contributions

Some entry level works, compared to those of the Muslim scientists, were made possible by the 13th century due to the availability of abundant translations of Muslim scientific treatises into the Romanic languages. Fibonacci popularized the Arabic numeral system in the Western world primarily through his composition in 1202 of Liber Abaci (Book of Calculation); and he introduced the sequence of numbers, which he used as an example in Liber Abaci, and later came to be known as Fibonacci sequence.

The book, Liber Abaci, showed the practical use and value of Arabic numerals including zero and a decimal, which he learnt from the translations of the Arabic treatise into Romanic, by applying the numerals to commercial bookkeeping, converting weights and measures, calculation of interest, money-changing, and other applications.

The book was well-received throughout educated Europe and had a profound impact on European thought. Europe was still using Roman numerals, ancient Egyptian multiplication method, and an abacus for calculations. Compared to these primitive methods, the use of Arabic numerals was a giant step forward. It represented a quantum advance in making business calculations easier and faster. This assisted the growth of banking and accounting in Europe.

The original 1202 manuscript is not known to exist. In a 1228 copy of the manuscript, the first section introduces the numeral system and compares it with others, such as Roman numerals, and methods to convert numbers to it. The second section explains uses in business, for example converting different currencies, and calculating profit and interest. Such calculations were important to the growing banking industry, and were highly cumbersome in the prevailing primitive methods.

The book also discusses irrational numbers and prime.

Liber Abaci posed and solved a problem involving the growth of a population of rabbits based on idealized assumptions. The solution, generation by generation, was a sequence of numbers later known as

Fibonacci numbers; although the sequence had been described by Indian mathematicians as early as the sixth century. In honesty, the sequence ought to be named as Hindu-Arabic sequence, rather than Fibonacci sequence.

Following is a list of books attributed to Fibonacci.

- Liber Abaci (1202), a book on calculations (English translation by Laurence Sigler, 2002).
- Practica Geometriae (1220), a compendium of techniques in surveying, the measurement and partition of areas and volumes, and other topics in practical geometry (English translation by Barnabas Hughes, Springer, 2008).
- Flos (1225), solutions to problems posed by Johannes of Palermo
- Liber quadratorum ("The Book of Squares") on Diophantine equations, dedicated to Emperor Frederick II.
- Di minor guisa (on commercial arithmetic; lost)
- Commentary on Book X of Euclid's Elements (lost)

As a cursory look at the already existing translations of the works of the Muslim scientists would easily show, most of Fibonacci's work is not original, rather, it is lifted off the existing works.

Biographical Summary

Fibonacci was born around 1170 to Guglielmo, an Italian merchant and customs official. Guglielmo directed a trading post in Bugia (Béjaïa) in modern-day Algeria), the capital of the Hammadid empire. Fibonacci travelled with him as a young boy, and it was in Bugia (Algeria) where he was educated.

Fibonacci travelled around the Mediterranean coast. He realized the advantages of the Arabic numerals system. Unlike the Roman numerals that the Europeans were using, the Arabic Numerals allowed easy calculation using a place-value system. In 1202, he completed the Liber Abaci (Book of Abacus or The Book of Calculation), which popularized Arabic numerals in Europe.

In 1240, the Republic of Pisa honored Fibonacci (referred to as Leonardo Bigollo) by granting him a salary in a decree that recognized him for the services that he had given to the city as an advisor on matters of accounting and instruction to citizens.

Fibonacci is thought to have died between 1240 and 1250, in Pisa.

54. Jordanus de Nemore

Jordanus de Nemore, also known as Jordanus Nemorarius and Giordano of Nemi,

(fl. 13th century),

was a thirteenth-century European mathematician and scientist.

The literal translation of Jordanus de Nemore (Giordano of Nemi) would indicate that he was an Italian.

Scientific Contributions

Some works were made possible by the 13th century due to the availability of abundant translations of Muslim scientific treatises into the Romanic languages. Jordanus de Nemore wrote on six mathematical topics: the science of weights; "algorismi" books on practical arithmetic; Arabic numerals arithmetic; algebra from Algebr wal Muqabela; geometry; and stereographic projection.

Following are Jordanus' works that have been published in critical editions in the twentieth century.

1. Mechanics: The three main treatises and the "Aliud commentum" version (Latin and English) are published in The Medieval Science of Weights, ed. Ernest A. Moody and Marshall Clagett (Madison: University of Wisconsin Press, 1952). The commentaries are also found in Joseph E. Brown,

"The 'Scientia de ponderibus' in the Later Middle Ages," PhD. Dissertation, University of Wisconsin, 1967.

The Liber de ponderibus and the "Aliud commentum" version were published by Petrus Apianus (= Peter Bienewitz) in Nuremberg, 1533; and the De ratione ponderis was published by Nicolò Tartaglia in Venice, 1565.

2. The Algorismi treatises: The articles by Gustaf Eneström, which contain the Latin text of the introductions, definitions and propositions, but only some of the proofs, were published in Biblioteca Mathematica, ser 3, vol. 7 (1906–07), 24-37; 8 (1907–08), 135-153; 13 (1912–13), 289-332; 14 (1913–14) 41-54 and 99-149.

3. Arithmetic (the De elementis arithmetice artis): Jacques Lefèvre d'Étaples (1455–1536) published a version (with his own demonstrations and comments) in Paris in 1496; this was reprinted, Paris, 1514. The modern edition is: H. L. L. Busard, Jordanus de Nemore, De elementis arithmetice artis. A Medieval Treatise on Number Theory (Stuttgart: Franz Steiner Verlag, 1991), 2 parts.

4. Algebra (De numeris data): The text was published in the 19th century, but a critical edition now exists: Jordanus de Nemore, De numeris datis, ed. Barnabas B. Hughes (Berkeley: University of California Press, 1981).

5. Geometry: "De triangulis" was first published by M.Curtze in "Mittheilungen des Copernicusvereins für Wissenschaft und Kunst" Heft VI - Thorn, 1887. See in Kujawsko-Pomorska Digital Library.

More recently, the Liber philotegni Iordani and the Liber de triangulis Iordani have been critically edited and translated in: Marshall Clagett, Archimedes in the Middle Ages (Philadelphia: American Philosophical Society, 1984), 5: 196-293 and 346-477, which is much improved over Curtze's edition.

6. Stereographic projection: The text of version 3 of Demonstratio de plana spera and the introduction were published in the sixteenth century – Basel, 1536 and Venice, 1558. All versions are edited and translated in: Ron B. Thomson, Jordanus de Nemore and the Mathematics of Astrolabes: De Plana Spera (Toronto: Pontifical Institute of Mediaeval Studies, 1978).

These content are amply available in the translated works of the works of the Muslim scientists. They are not the original research of Jordanus. In fact, the topics discussed by Jordanus are rather shallow compared with the sophistication and depth of the original sources.

Biographical Summary

We know nothing about Jordanus personally, other than the approximate date of his work.

Cited in the early manuscripts simply as "Jordanus", he was later given the sobriquet of "de Nemore" ("of the Forest," "Forester") which does not add any firm biographical information. In the Renaissance his name was often given as "Jordanus Nemorarius", an improper form.

It is assumed that Jordanus did work in the first part of the thirteenth century (or even in the late twelfth) since his works are contained in a booklist, the Biblionomia of Richard de Fournival, compiled between 1246 and 1260.

55. Al-Urdi

Moayad Al-Din Al-Urdi Al-Amiri Al-Dimashqi

(Arabic: مؤيد الدين العرضي العامري الدمشقي),

(d. 1266),

was an astronomer and geometer from Syria.

Scientific Contributions

Al-Urdi's most notable works are Risālat al-Raṣd, a treatise on observational instruments, and Kitāb al-Hayʾa (كتاب الهيئة), a work on theoretical astronomy.

His influence can be seen on Bar Hebraeus and Qutb al-Din al-Shirazi, in addition to being quoted by Ibn al-Shatir.

Al-Urdi contributed to the construction of the observatory outside of the city, constructing special devices and water wheels in order to supply the observatory, which was built on a hill, with drinking water.

He also constructed some of the instruments used in the observatory, in the year 1261/2. Al-Urdi's son, who also worked in the observatory, made a copy of his father's Kitāb al-Hayʾa and also constructed a celestial map in 1279.

Al-Urdi is a member of the group of Islamic astronomers of the 13th and 14th centuries who were active in the criticism of the astronomical model presented in Ptolemy's Almagest.

Saliba (1979) showed Bodleian ms. Marsh 621 as a copy of Al-Urdi's Kitāb al-Hayʾa, based on which he argued that Al-Urdi's contributions predated Al-Tusi.

Otto E. Neugebauer in 1957 argued that the works of this group of astronomers, perhaps via Ibn al-Shatir, must have been received in 15th-century Europe and ultimately influenced the works of Copernicus.

Of special significance is the "Urdi lemma" in particular, which generalizes Apollonius' theorem. This generalization has significance in that it *allows an equant in an astronomic model to be replaced with an equivalent epicycle* that moved around a deferent centered at half the distance to the equant point.

Biographical Summary

Al-Urdi was Born circa 1200, presumably, from the nisba al-ʿUrḍī, in the village of ʿUrḍ in the Syrian desert between Palmyra and Resafa. He came to Damascus at some point before 1239, where he worked as an engineer and teacher of geometry. He built instruments for al-Malik al-Mansur of Hims.

In 1259 he moved to Maragha in northeastern Iran, after being asked by Nasir al-Din al-Tusi to help establish the Maragha observatory under the patronage of Hulagu.

Al-Urdi died in 1266 AD.

56. Ibn 'Adlan

'Afīf al-Dīn 'Alī ibn 'Adlān al-Mawsilī

(Arabic: عفيف لدين علي بن عدلان الموصلي),

(1187–1268 CE),

born in Mosul, was a cryptologist, linguist and poet who is known for his early contributions to cryptanalysis, to which he dedicated at least two books.

Scientific Contributions

Ibn 'Adlān wrote his Cryptanalysis work containing rules of cryptanalysis. He detailed a set of twenty "rules", discussing practical details. Among Ibn 'Adlan's original contributions were methods for breaking no-space monoalphabetic cryptograms, a type of ciphers which were developed to evade analysis techniques described earlier by Al-Kindi.

In his treatise Ibn 'Adlan includes a real-life example of a cryptogram that he deciphered and his full process in breaking it: demonstrating an authentic experience of a highly skilled cryptanalyst.

Early Arabic bibliographies attributed three titles to Ibn 'Adlan, including one on cryptanalysis, "Fi hall al-mutarjam" (On Cryptanalysis), also known as "Al-mu'allaf lil-malik al-'Ashraf" (The [book] written for King al-Ashraf).

In addition, a reference in “Fi hall al-mutarjam” points to another book, Al-Mu'lam (The told [book]), which is now lost, in which he describes algorithms for analyzing cryptograms. His other two works were titled “Al-Intihab li-kashf al-'abyat al-mushkilat al-i'rab” and “'Uqlat al-mujtaz fi hall al-aljaz”.

While encryption of messages is an older practice, the Arabic Scientists originated the science of decrypting an encrypted message. The earliest surviving work found on the topic of cryptanalysis is the “Risalah fi Istikhraj al-Mu'amma” ("Treatise on Deciphering Cryptographic Messages") written by Al-Kindi (c. 801–873). Reports are also found on other works before al-Kindi, among the earliest of which is al-Mu'amma ("The Book of Cryptographic Messages"), written by al-Khalil ibn Ahmad in the 8th century, but they are now lost.

“Fi hall al-mutarjam” was like a handbook or a manual, describing Ibn 'Adlan's twenty "rules" or techniques of cryptanalysis, grouped into nine themes. Unlike the cryptological treatises of Al-Kindi before him and later Ibn al-Durayhim (c. 1312–1361), which provide theoretical background on cryptography including systematic explanations on types of ciphers, Ibn 'Adlan's treatise focuses on the practical matters and specific methods in breaking encrypted texts, often in a more detailed manner than Al-Kindi. The work's introduction section does include a brief description of the simple substitution encipherment method, and encourages its readers to read other sources to learn about other methods.

One of Ibn 'Adlan's most original contribution in this treatise is the cryptanalysis of no-space monoalphabetic cryptograms (al-mudmaj)—encrypted texts that do not include a space to denote separation between words. This type of cryptograms was not mentioned by al-Kindi: it was developed by subsequent cryptographers (code makers). According to ibn 'Adlan, the cryptographers of his time allege that their ciphers using the no-space method can defy detection and analysis. Ibn 'Adlan also wrote on the analysis of ciphers in which the space is represented by variable symbols.

In the west, this type of cryptanalysis was only attested in the sixteenth century in the works of the Italian Giambattista della Porta — way after the works of the Muslim scientists had become widely available through their translations into Latin languages.

Frequency analysis technology

Ibn 'Adlan recommended the use of frequency analysis approach. He also uses the analysis of consecutive letters based on knowing how many times each letter can possibly occur consecutively in Arabic sentences and the specific ways they can do so.

"Fi hall al-mutarjam" also deals with frequency analysis: Ibn 'Adlan follows al-Kindi's data on the frequency of Arabic letters, the numbers provided by the two authors are identical, and divided the Arabic alphabet into seven common (frequently-occurring), eleven medium, and ten rare letters. Ibn 'Adlan also presents a table of the most common two or three letter words.

Ibn 'Adlan analyzed a minimum sample size, a lower limit of text length that can be cryptanalyzed using its frequency of letters. This lower limit is about 90 characters, approximately three times the length of the Arabic alphabet. Below this limit, according to Ibn 'Adlan the occurrence of letters will not follow the provided frequency distribution.

This conclusion is consistent with modern study of statistical distributions.

The treatise includes the cryptanalysis of common adjacent letters, the Arabic definite article ال (al-, 'the'), and letters frequently occurring at the beginning or the end of a word.

Ibn 'Adlan writes on the probable words in the opening and closing section of a text (such as the Arabic formula Bismillah, "In the name of God"). He adds special principles for analyzing encrypted poetry, including the knowledge of prosody, rhymes and meters. He then explains his cryptanalysis steps, moving from the ciphertext to possible solutions, then to the suspected, the probable, and eventually the confirmed solution.

In the closing section of the book, Ibn 'Adlan includes a real-life example of a cryptogram that he broke and his full process in deciphering it, including his false starts, thought process, and eventual solution. This demonstration of the decrypting process is intriguing and provides an authentic experience of a highly skilled cryptanalyst.

A copy of "Fi hall al-mutarjam" is preserved in the library of the Süleymaniye Mosque of Istanbul (manuscript number 5359). A modern edition was prepared by editors Muhammad Mrayati, Yahya Meer Alam and Hassan al-Tayyan and published by the Arab Academy of Damascus in 1987, including introductions and explanatory materials from the editors. It was translated into English in 2004.

Biographical Summary

'Afif al-Din 'Ali ibn 'Adlan was born in Mosul in 1187 AD). He received education in Baghdad, including lessons on syntax by the grammarian Abu al-Baqa al-Ukbari. Subsequently, he lived in Damascus for a time. He became a teacher of the Arabic language at the Al-Salihiyya Mosque of Cairo until his death in 1268 CE.

In addition to writing treatises on linguistics and cryptanalysis, he was considered an authority in literature and wrote poems himself.

He was in contact with various rulers of his time, and in this capacity, he gained practical experience in cryptanalysis or the science of making and breaking encoded messages (hall al-mutarjam). He dedicated his work "On Cryptanalysis", his only surviving work on the topic, to Al-Ashraf Musa (r. 1229–1237), the Ayyubid Emir of Damascus.

He wrote three other books, including Al-Mu'lam (The Told), also on cryptanalysis, but it is now lost.

He was known by multiple nisbas: al-Mawsili (of Mosul), al-Nahwi (the Grammarian) and al-Mutarjim (the Cryptoanalyst).

57. Nasir al-Din al-Tusi

Muhammad ibn Muhammad ibn al-Hasan al-Tūsī

(Persian: محمد ابن محمد ابن حسن طوسی),

(18 February 1201 – 26 June 1274),

was a Persian polymath, architect, philosopher, physician, scientist, and theologian.

Scientific Contributions

Tusi has about 150 works, of which 25 are in Persian and the remaining are in Arabic, and there is one treatise in Persian, Arabic and Turkish.

Following are some of his treatises.

- A Treatise on the Astrolabe by Tusi, Isfahan 1505
- Sayr wa-Suluk (The Voyage) - Autobiography
- Kitāb al-Shakl al-qattāʿ Book on the complete quadrilateral. A five-volume summary of trigonometry.
- Al-Tadhkirah fi'ilm al-hay'ah – A memoir on the science of astronomy. Many commentaries were written about this work called Sharh al-Tadhkirah (A Commentary on al-Tadhkirah) - Commentaries were written by Abd al-Ali ibn Muhammad ibn al-Husayn al-Birjandi and by Nazzam Nishapuri.
- Akhlaq-i Nasiri – A work on ethics.
- al-Risalah al-Asturlabiyah – A Treatise on the astrolabe.

- Zij-i Ilkhani (Ilkhanic Tables) – A major astronomical treatise, completed in 1272.
- Sharh al-Isharat (Commentary on Avicenna's Isharat)
- Awsaf al-Ashraf a short mystical-ethical work in Persian.
- Tajrīd al-I'tiqād (Summation of Belief) – A commentary on Shia doctrines.
- Talkhis al-Muhassal (summary of summaries).
- Maṭlūb al-mu'minīn (Desideratum of the Faithful)
- Aghaz u anjam - Esoteric interpretation of the Quran.

During his stay in Nishapur, Tusi established a reputation as an exceptional scholar. Tusi's prose writing, which numbers over 150 works, represent one of the largest collections by a single author. Writing in both Arabic and Persian, Nasir al-Din Tusi dealt with both religious topics and scientific subjects.

Astronomy

In astronomy, Tusi convinced Hulegu Khan to construct an observatory for establishing accurate astronomical tables for better astrological predictions. Beginning in 1259, the Rasad Khaneh observatory was constructed in Azarbaijan, south of the river Aras, and to the west of Maragheh, the capital of the Ilkhanate Empire.

Based on the observations in this, being the most advanced observatory in the world, Tusi made very accurate tables of planetary movements as depicted in his book Zij-i ilkhani (Ilkhanic Tables). This book contains astronomical tables for calculating the positions of the planets and the names of the stars. His model for the planetary system was the most advanced in the world, and was used extensively.

His famous student Shams al-Din al-Bukhari was the teacher of Byzantine scholar Gregory Chioniades, who had in turn trained astronomer Manuel Bryennios in about 1300 AD in Constantinople.

For his planetary models, he invented a geometrical technique called a Tusi-couple, which *generates linear motion from the sum of two circular motions.* He used this technique to replace Ptolemy's problematic equant for many planets, but was unable to find a solution to Mercury, which was solved later by Ibn al-Shatir as well as Ali Qushji. The Tusi couple was later employed in Ibn al-Shatir's geocentric model and Nicolaus Copernicus' heliocentric Copernican model.

He also calculated the value for the annual precession of the equinoxes and contributed to the construction and usage of some astronomical instruments including the astrolabe.

Ṭūsī criticized Ptolemy's use of observational evidence to show that the Earth was at rest, noting that such proofs were not decisive. Tusi's arguments in criticisms of Ptolemy were later used by Copernicus in 1543 to defend the Earth's rotation.

About the real essence of the Milky Way, Ṭūsī in his Tadhkira writes: "The Milky Way, i.e., the galaxy, is made up of a very large number of small, tightly-clustered stars, which, on account of their concentration and smallness, seem to be cloudy patches. because of this, it was likened to milk in color."

This ansatz was proven, three centuries later in 1610, when Galileo Galilei used a telescope to study the Milky Way and discovered that it is really composed of a huge number of faint stars.

However, the credit is attributed entirely to Galileo.

Logic

In logic, Nasir al-Din Tusi was a supporter of Avicennian logic, and wrote the following commentary on Avicenna's theory of absolute propositions:

"What spurred him to this was that, in the assertoric syllogistic, Aristotle and others sometimes used contradictories of absolute propositions on the assumption that they are absolute; and that was why so many decided that absolutes did contradict absolutes.

When Avicenna had shown this to be wrong, he wanted to develop a method of construing those examples from Aristotle."

Mathematics

In mathematics, Al-Tusi was the first to write a work on trigonometry independently of astronomy. Al-Tusi, in his Treatise on the Quadrilateral, gave an extensive exposition of spherical trigonometry, distinct from astronomy. It was in the works of Al-Tusi that trigonometry achieved the status of an independent branch of pure mathematics distinct from astronomy, to which it had been linked for so long.

He was the first to list the six distinct cases of a right triangle in spherical trigonometry.

In his "On the Sector Figure", appears the famous Sine Law for plane triangles.

(a/Sin A) = (b/Sin B) = (c/Sin C)

Where A, B, C are angles of the triangle, and a, b, c are the lengths of the sides opposite these angles.

He also stated the sine law for spherical triangles, discovered the law of tangents for spherical triangles, and provided proofs for these laws.

Color Theory

In color theory, while Aristotle (d. 322 BCE) had suggested that all colors can be aligned on a single line from black to white, Ibn-Sina (d. 1037) described that there were three paths from black to white, one path via grey, a second path via red and the third path via green. Al-Tusi (ca. 1258) stated that there are no less than five of such paths, via lemon (yellow), blood (red), pistachio (green), indigo (blue) and grey.

This text, which was copied in the Middle East numerous times until at least the nineteenth century as part of the textbook "Revision of the Optics" (Tanqih al-Manazir) by Kamal al-Din al-Farisi (d. 1320), made color space effectively two-dimensional.

Robert Grosseteste (d. 1253) proposed an effectively three-dimensional model of color space.

Biology

In biology, Tusi wrote in his Akhlaq-i Nasiri about several biological topics. He described "grasses which grow without sowing or cultivation, by the mere mingling of elements," as closest to minerals. Among plants, he considered the date-palm as the most highly developed, since "it only lacks one thing further to reach (the stage of) an animal: to tear itself loose from the soil and to move away in the quest for nourishment."

The lowest animals "are adjacent to the region of plants: such are those animals which propagate like grass, being incapable of mating [...], e.g. earthworms, and certain insects". The animals "which reach the stage of perfection [...] are distinguished by fully developed weapons", such as antlers, horns, teeth, and claws.

Tusi *described these organs as adaptations to each specie's lifestyle,* (in a way anticipating natural theology). This is akin to a postulate of the theory of evolution attributed to Darwin.

He continued: "The noblest of the species is that one whose sagacity and perception is such that it accepts discipline and instruction: thus, there accrues to it the perfection not originally created in it. Such are the schooled horse and the trained falcon. The greater this faculty grows in it, the more surpassing its rank, until a point is reached where the (mere) observation of action suffices as instruction: thus, when they see a thing, they perform the like of it by mimicry, without training [...]. This is the utmost of the animal degrees, and the first of the degrees of Man in contiguous therewith."

Thus, in this paragraph, Tusi described different types of learning, recognizing observational learning as the most advanced form, and correctly attributing it to certain animals.

Tusi seems to have perceived man as belonging to the animals, since he stated that "the Animal Soul [comprising the faculties of perception and movement ...] is restricted to individuals of the animal species", and that, by possessing a "Human Soul, [...] mankind is distinguished and particularized among other animals."

Some scholars have interpreted Tusi's biological writings as suggesting that he adhered to some kind of *evolutionary theory.* However, theory of evolution is entirely attributed to Darwin.

Chemistry

Tusi contributed to the field of chemistry, stating an early *law of conservation of mass.*

In philosophy, Tusi contributed many writings to the topic of philosophy. Amongst his philosophical work are his disagreements with fellow philosopher Avicenna. His most famous philosophical work is Akhlaq-i nasiri or Nasirean Ethics. Within this work he discusses and compares Islamic teachings to the ethics of Aristotle and Plato.

Tusi's book became a popular ethical work in the Muslim world, specifically in India and Persia. Tusi's work also left an impact on

Shi'ite Islamic theology. His book Targid also called Catharsis is significant in Shi'ite theology. He also contributed five works to the subject of logic; which were highly regarded by his contemporaries, but also achieved some notoriety.

Some scholars believe that Nicolaus Copernicus was influenced by Middle Eastern astronomers due to uncanny similarities between his work and the uncited work of these Islamic scholars, including Nasir al-Din al-Tusi, Ibn al-Shatir, Muayyad al-Din al-Urdi, and Qutb al-Din al-Shirazi, and al-Tusi specifically. There are unmistakable similarities in the Tusi couple and Copernicus' geometric method of removing the Equant from mathematical astronomy.

> Not only do both of the methods match geometrically, more importantly, they both use the same exact lettering system for each *vertex*; a detail that seems too preternatural to be happenstance. Moreover, the fact that several other details of his model also mirror other Islamic scholars bolsters the notion that Copernicus' work may not have been his own.

There is strong evidence that the mathematics and theories on the work of Nasir al-Din al-Tusi did make the journey to Europe: the Islamic school in Maragheh, which was home to Nasir al-Din al-Tusi's observatory in Muslim Spain, was only a stone's throw away; translations of the works into European languages did exist, such as the Greek translations by Gregory Choniades; and there was enough traffic of scholars from Islamic centers to Europe. On the other hand, those who like to refute this evidence do so on general grounds which adds up to little more than hand waving.

Biographical Summary

Nasir al-Din al-Tusi was born in the city of Tus in Khorasan (northeastern Iran) in the year 1201 and began his studies at an early age, as is customary among Muslims: in Hamadan and Tus he studied the Quran, hadith, jurisprudence, logic, philosophy, mathematics, medicine, and astronomy.

He was born into a Shī'ah family and lost his father at a young age. Fulfilling the wish of his father, the young man took learning and scholarship very seriously and traveled far and wide to attend the lectures of renowned scholars and acquired knowledge; an exercise highly encouraged in his Islamic faith.

At a young age, he moved to Nishapur to study philosophy under Farid al-Din Damad and mathematics under Muhammad Hasib. He met also Attar of Nishapur, the legendary Sufi master who was later killed by the Mongols. He attended the lectures of Qutb al-Din al-Misri.

Nasir-al-Din Tusi writes in his work, Maṭlūb al-muʾminīn, (Desideratum of the Faithful)

> "To become people of spiritual reality, it is incumbent to fulfill the symbolic elucidation (ta'wīl) of the seven pillars of the religious law (sharīʿat)".

He also explains that fulfilling the religious law is much easier than fulfilling its spiritual interpretation.

He explains in his book "Aghaz u anjam" that the sacred accounts of history that we perceive within the bounds of space and time symbolize events that have no such restrictions. They are only expressed in this way so that humans are able to comprehend them.

In Mosul, al-Tusi studied mathematics and astronomy with Kamal al-Din Yunus (d. AD 1242), a pupil of Sharaf al-Dīn al-Ṭūsī.

Later on, he corresponded with Sadr al-Din al-Qunawi, the son-in-law of Ibn Arabi, and it seems that *mysticism, as propagated by Sufi masters of his time, was not appealing to him.*

Once the occasion was suitable, he composed his own manual of philosophical Sufism in the form of a small booklet entitled "Awsaf al-Ashraf" (The Attributes of the Illustrious).

As the armies of Genghis Khan swept his homeland, he was employed by the Nizari Ismaili state and, while moving from stronghold to stronghold, made his most important contributions in science. First, he was in those of the Quhistan region under Muhtasham Nasir al-Din Abd al-Rahim ibn Abi Mansur, where he wrote the Nasirean Ethics. He was later sent to the major castles of Alamut and Maymun-Diz to continue his career under Nizari Imam Ala al-Din Muhammad. He was captured after the fall of Maymun-Diz to the Mongol forces under Hulagu Khan. Nasir al-Din Tusi explains in his "Sayr wa-Suluk" a literary devastation such as the devastation of the Alamūt libraries in 1256 would not waver the spirit of the Nizari Ismaili community because they give more importance to the "living book" (the Imam of the Time) rather than the "written word". Their hearts are attached to the Commander of the Believers (amir al-mu'minin), not just the "command" itself. There is always a present living Imam in the world, and following him, a believer will never go astray.

Nasir al-Din al-Tusi died in 1274 AD. Following poem of his is a good advice:

Anyone who knows, and knows that he knows,
makes the steed of intelligence leap over the vault of heaven.
Anyone who does not know but knows that he does not know,
can bring his lame little donkey to the destination nonetheless.
Anyone who does not know, and does not know that he does not know,
is stuck forever in double ignorance.

58. Hasan al-Rammah

Hasan al-Rammah

(Arabic: حسن الرماح),

(died 1295),

was a chemist and engineer from Syria during the Mamluk Sultanate.

Scientific Contributions

Hasan al-Rammah studied the science of gunpowder and explosives; designed instruments of warfare; and he invented the first torpedo in the World.

Al-Rammah called his early torpedo "an egg which moves itself and burns." It was made of two sheet-pans of metal fastened together and filled with naptha, metal filings, and saltpeter. It was intended to move across the surface of the water, propelled by a large rocket and kept on course by a small rudder.

Al-Rammah devised several new types of gunpowder, and he invented a new type of fuse and two types of igniters.

Biographical Summary

Hasan al-Rammah was a chemist and a chemical engineer from Syria. He died in 1295 AD.

59. Al-Ashraf Umar II

Al-Malik Al-Ashraf (Mumahhid Al-Din) Umar Ibn Yūsuf Ibn Umar Ibn Alī Ibn Rasul

(Arabic: عمر بن يوسف بن عمر بن علي بن رسول الغساني), also known as Umar Ibn Yusuf and Al-Asharaf Umar II,

(1242 – 1296 AD)

was a mathematician, astronomer and physician; and also the third Rasulid sultan.

Scientific Contributions

He is known for writing the first description of the use of a magnetic compass for determining the direction in navigation. His works on astronomy contain important information on earlier sources.

In a treatise about astrolabes and sundials, al-Ashraf includes several paragraphs on the construction of a compass bowl (ṭāsa). He then uses the compass to determine the north point, the meridian (khaṭṭ niṣf al-nahār), and the Qibla towards Mecca. This is the first mention of a compass in a scientific text.

Biographical Summary

Umar Ibn Yusuf was the third Sultan in Yemen under Rasulid dynasty (1229 AD to 1454). He was born in 1242 AD in Yemen and he died in Yemen in 1296.

60. Zayn al-Din al-Amidi

Zayn al-Din 'Ali ibn Ahmad al-Amidi

(Arabic: زين الدين علي بن أحمد الآمدي),

(died 1312 AD),

was a blind inventor of technology for blind people to read books.

Scientific Contributions

Zayn al-Din al-Amidi invented a system to study and recognize books. His method involved the use of fruit stones as a reading means for the blind.

Zayn al-Din al-Amidi was blind but he wrote a book "Nakt al-Himyan fi Nukat al-'Umyan" (Emptying the pockets for anecdotes about blind people).

He invented a technology for blind people to read books. Al Amidi was five centuries before his time, as the Braille process was reinvented in Europe in 1824 AD, similarly by a blind person named Louis Braille.

Zayn al-Din al-Amidi was very knowledgeable, and he was a trader in books. He could pick out the desired volume, touch the book and determine the number of its pages; he would touch the page and determine how many lines it had, the type of script and its color. Naturally, he knew the prices of the books.

Biographical Summary

Zayn al-Din al-Amidi was a blind Arab who invented the science for the blind to read. He died in 1312 AD.

61. Ibn al-Raqqam

Ibn Al-Raqqam Muḥammad Ibn Ibrahim Al-Mursi Al-Andalusi Al-Tunisi Al-Awsi

(Arabic: ابن الرقام الأوسي),

(1250–1315), also known as Ibn Al-Raqqam,

was a 13th century astronomer, mathematician and physician; but also, a Sunni Muslim theologian and jurist, from Andalusia.

Scientific Contributions

Although several works have been attributed to him by Ibn Al-Khatib, only three ones have survived in an extant form. Two of his works are astronomical tables. The tables were created to adapt the coordinates of two different cities, Béjaïa and Tunis. The third work, "Risāla fi ʿilm Al-Zilal", is an important treatise on sundials, and the only complete one of its kind to have survived from Al-Andalus.

In Astronomy Ibn Al-Raqqam wrote "Arnau Teruel", it was translated as *Padre y Dios del Nuevo Mundo;* "Risāla fī ʿilm al-ẓilāl", there is a copy of it in the first Escorial No. (7/913) and the second number (12/918); "Al-Zīj al-qawīm fī funūn al-taʿdīl wa-ʾl-taqwīm", there is a copy of it in the public library in Rabat, number (260); "Taedil munakh al'ahlat"; and "Al-Zīj Al-Mustawfi".

In Medicine Ibn Al-Raqqam wrote "The Great Book", "Kitāb al-Ḥayawān wa-ʾl-khawāṣṣ" (The Book of Animals and Properties); "A summary of competence (or abbreviation) in the knowledge of powers and properties"; "Treating diseases"; "Authorship in Medicine" which consists of two parts, and there is a copy of it in the public treasury in Rabat, number (2667).

In Jurisprudence Ibn Al-Raqqam wrote "Abkār al-afkār fī al-uṣūl', and "Talkhis almubahath".

In Mathematics Ibn Al-Raqqam wrote "Al-Tanabih waltabsir fi qawaeid altksi", there is a copy of it in the Hassaniya Treasury in Rabat, No. (4749).

In Agriculture Ibn Al-Raqqam wrote on "Plants".

Biographical Summary

Abu Abdullah Ibn Al-Raqam was born in Murcia in 1250, in a family with the nisba al-Awsi, probably from the Banu Aws tribe, and grew up and learned there until the city was annexed by Castile in 1266. He left Murcia for the city of Bejaia, in present-day Algeria, and lived there until he went to Tunisia and spent time there writing some of his books. Later in his life, he settled in Granada, the capital of the Emirate of Granada, after accepting an invitation from Muhammad II of Granada.

Remarks

Ibn al-Raqqam is one of the polymaths who could have been included as a Natural Scientist, a Medical Scientist, or a Religious Scientist. We chose to include him as a Natural Scientist, and will not include him also in other categories because of our policy to count a scientist only in one category.

62. Ibn al-Banna' al-Marrakushi

Ibn al-Bannāʾ al-Marrākushī,

(Arabic: ابن البنّاء), also known as Abu'l-Abbas Ahmad ibn Muhammad ibn Uthman al-Azdi,

(29 December 1256 in Marrakush – c. 1321),

was a mathematician, astronomer, Islamic scholar, and Sufi.

Scientific Contributions

Ibn al-Banna' wrote between 51 and 74 treatises, encompassing such varied topics as Algebra, Astronomy, Linguistics, Rhetoric, and Logic. One of his works, called Talkhīṣ ʿamal al-ḥisāb (Arabic: تلخيص أعمال الحساب) (Summary of arithmetical operations), includes topics such as fractions and sums of squares and cubes. Another, called Tanbīh al-Albāb, covers topics related to:

calculations regarding the drop in irrigation canal levels,

arithmetical explanation of the Muslim laws of inheritance,

determination of the hour of the Asr prayer,

explanation of frauds linked to instruments of measurement,

enumeration of delayed prayers which have to be said in a precise order, and

calculation of legal tax in the case of a delayed payment.

He wrote Rafʿ al-Ḥijāb (Lifting the Veil) which covered topics such as computing square roots of a number and the theory of continued fractions. This was the first known mathematical work to use an algebraic approach, further developed by Abū al-Hasan ibn Alī al-Qalasādī two centuries later.

Ibn al-Banna' wrote Jami‘ al-Mabadi’ wa’l-Ghayat (Collection of the Principles and Objectives in the Science of Timekeeping), a comprehensive work on astronomy.

Biographical Summary

Ibn al-Banna' (lit. the son of the architect) was born in Marrakesh in 1256; he is named al-Marrākushī after that city. Having learned mathematical and geometrical skills, he translated Euclid's Elements into Arabic.

63. Ibn al-Durayhim

Alī ibn Muḥammad Ibn al-Durayhim

(Arabic: علي بن محمد ابن الدريهم),

(1312–1359/62 AD),

was a cryptologist who gave detailed research results of eight cipher systems that discussed substitution ciphers, leading to the earliest suggestion of a "tableau" of the kind that two centuries later became known as the "Vigenère table".

Scientific Contributions

His book entitled "Clear Chapters Goals and Solving Ciphers" (مقاصد الفصول المترجمة عن حل الترجمة) was recently discovered, but has yet to be published. It includes the use of the statistical techniques pioneered by Al-Kindi and Ibn 'Adlan.

The cryptology of Ibn al-Durayhim gave detailed descriptions of eight cipher systems that discussed substitution ciphers, leading to the earliest suggestion of a "tableau". Two centuries later it became known as the "Vigenère table".

Biographical Summary

Ibn al-Durayhim was born in 1312 AD and he died in 1359/62.

64. Ibn al-Shatir

ʿAbu al-Ḥasan Alāʾ al-Dīn ʿAlī ibn Ibrāhīm al-Ansari known as Ibn al-Shatir or Ibn ash-Shatir

(Arabic: ابن الشاطر),

1304–1375),

was an astronomer, mathematician and engineer. He worked as muwaqqit (موقت, religious timekeeper) in the Umayyad Mosque in Damascus and constructed a sundial for its minaret in 1371/72.

Scientific Contributions

Astronomy:

Ibn al-Shatir most important astronomical treatise was "kitab nihayat al-sul fi tashih al-usul" ("The Final Quest Concerning the Rectification of Principles"). In it he drastically reformed the Ptolemaic models of the Sun, Moon and planets. His model incorporated the mathematics of the Urdi lemma, and eliminated the need for an equant (a point on the opposite side of the center of the larger circle from the Earth) by introducing an extra epicycle (the Tusi-couple). This departed from the Ptolemaic system in a way that was mathematically identical (but conceptually very different) to what Nicolaus Copernicus did in the 16th century. This new planetary model was published in his work the al-Zij al-jadid (The New Planetary Handbook.) Before the "kitab nihayat al-sul fi tashih al-usul" was written, there was a treatise that Ibn al-Shatir wrote which described the observations and procedures that led him to create his new planetary models.

Unlike previous astronomers before him, Ibn al-Shatir was not concerned with adhering to the so called "principles of natural philosophy" by Aristotelian. Rather, he produced a model that was more consistent with observations and contemporary concepts. For example, it was Ibn al-Shatir's concern for observational accuracy which led him to eliminate the epicycle in the Ptolemaic solar model and all the eccentrics, epicycles and equant in the Ptolemaic lunar model. Shatir's new planetary model consisted of epicycles instead of equant. His model was thus in better agreement with observations than any previous model, and was also the first that permitted empirical testing.

Ibn al-Shatir's work marked a turning point in astronomy, which may be considered a "Scientific Revolution".

Ibn al-Shatir's model for the appearances of Mercury, showed the multiplication of epicycles.

Ibn al-Shatir's Solar Model exemplifies his commitment towards accurate observational data. Ibn al-Shatir in fact established a fundamental scientific paradigm:

The experiments are of primary significance and can render a theory into questionable validity.

Using this overriding principle, Ibn al-Shatir discarded both "principles of natural philosophy" by Aristotelian, and the models of Ptolemy.

Ibn al-Shatir's model created a new eccentricity for the solar model. And with his numerous observations, Ibn al-Shatir was able to generate a new maximum solar equation (2;2,6°), which he found to have occurred at the mean longitude λ 97° or 263° from the apogee. He identified that 7;7 and 2;7 were the radii of the epicycles. In addition, his final results for apparent size of the solar diameter were concluded to be at apogee (0;29,5), at perigee (0;36,55), and at mean distance (0;32.32). The longitude of the planets was defined as a function of the mean longitude and the anomaly. Rather than calculating every possible value, which would be difficult and labor-intensive, four functions of a single value were calculated for each planet and combined to calculate quite accurately the true longitude of each planet.

To calculate the true longitude of the moon, Ibn al-Shatir assigned two variables, η, which represented the moon's mean elongation from the sun, and γ, which represented its mean anomaly. To any pair of these values was a corresponding e, or equation which was added to the mean longitude to calculate the true longitude. Ibn al-Shatir used the same mathematical scheme when finding the true longitudes of the planets, except for the planets the variables became α, the mean longitude measured from apogee (or the mean center) and γ which was the mean anomaly as for the moon. A correcting function c3' was tabulated and added to the mean anomaly γ to determine the true anomaly γ'.

As shown in Shatir's model, it was later discovered that Shatir's lunar model had a very similar concept as Copernicus.

Although Ibn al-Shatir's system was geocentric, the mathematical details of his system were identical to those in Copernicus's De revolutionibus. Furthermore, the exact replacement of the equant by two epicycles used by Copernicus in the Commentariolus was the same as the work of Ibn al-Shatir, one century earlier. Ibn al-Shatir's lunar and Mercury models are also identical to those of Copernicus. However, the accreditation goes entirely to Copernicus.

However, Copernicus's Mercury model was flawed because he was not able to properly understand the model created by Ibn al-Shatir, a century earlier.

Copernicus also translated Ptolemy's geometric models to longitudinal tables in the same way as Ibn al Shatir did when constructing his solar model.

This body of similarity has led some scientists to argue that Copernicus must have used the work of ibn al-Shatir.

A Byzantine manuscript containing a solar model diagram with a second epicycle, and other works of the Muslim scientists, was discovered to have been in Italy at the time of Copernicus: provides evidence of transmission of astronomical theories from the Muslim scientists to the newly initiated European scientists.

The Greek had sundials too, but they had nodus-based with straight hour lines which meant that the hours in the day would be unequal (temporary hours) depending on the season. Each day was split into twelve equal segments which meant that the hours would have been shorter in the winter and longer in the summer due to the activity of the sun. Ibn al-Shatir was aware that "using a gnomon that is parallel to the Earth's axis will produce sundials whose hour lines indicate equal hours on any day of the year." His sundial is the oldest polar-axis sundial still in existence.

The concept later appeared in European sundials from 1446.

The idea of using hours of equal time length throughout the year was the innovation of Ibn al-Shatir in 1371, based on earlier developments in trigonometry by al-Battānī. Before the Islamic scholar created the improved sundial, he had to understand the sundial created by his predecessors.

Ibn al-Shatir also invented a timekeeping device called "Sandūq al-Yawāqīt li maʿrifat al-Mawāqīt" (jewel box), which incorporates both a universal sundial and a magnetic compass. He invented it for the purpose of finding the times of prayers. The "Sandūq al-Yawāqīt li maʿrifat al-Mawāqīt" had a moveable hole in it which allowed the user to find the hour angle of the sun. If this angle was suitable with the horizon, then the user could use it as a polar sundial. This device is preserved in the museum of Aleppo.

Ibn al-Shatir created a sundial which was placed on top of the Madhanat al-Arus (The Minaret of the Bride) in the Umayyad Mosque. The sundial was created on a slab of marble which was approximately 2 meters by 1 meter. The sundial being engraved on the marble was so that Ibn al-Shatir could read the time of the day in equinoctial (equal times) hours for the prayer times. This sundial was later removed in the eighteenth century and a replica was put in its place. The original sundial was placed in the Damascus archeology museum.

Ibn al-Shatir created another sundial but in smaller dimensions (12 cm x 12 cm × 3 cm) to find out the prayer times of midday and the afternoon. This sundial was able to tell the local meridian and the direction of the mecca. Other notable instruments invented by Ibn al-Shatir include a reversed astrolabe and an astrolabic clock. The astrolabe that he created was called the al-āla al-jāmiʿa (the universal instrument). This astrolabe was created by Ibn al-Shatir when he wrote on the ordinary planispheric astrolabe and when he wrote on the two most common quadrants (the astrolabic and the trigonometric varieties). These two common quadrants were modified versions of the sine quadrant.

He also created a set of tables that had values of spherical astronomical functions for prayer times. The tables displayed the times for the morning, afternoon, and evening prayers. The latitude that was used to create the table was 34° (which corresponded to a location north of Damascus).

Biographical Summary

Ibn al-Shatir was born in Damascus, Syria, around the year 1304. His father passed away when he was six years old. His grandfather took him in which resulted in al-Shatir learning the craft of inlaying ivory. Ibn al-Shatir traveled to Cairo and Alexandria to study astronomy, which inspired him. After completing his studies with Abu 'Ali al-Marrakushi, Ibn al-Shatir returned to his home in Damascus where he was then appointed muwaqqit (timekeeper) of the Umayyad Mosque.

Part of his duties as muqaqqit involved keeping track of the times of the five daily prayers and when the month of Ramadan would begin and end. To accomplish this, he created a variety of astronomical instruments. He made several astronomical observations and calculations both for the purposes of the mosque, and to fuel his research.

These observations and calculations were organized in a series of astronomical tables. His first set of tables, which have been lost over time, contained entries on the Sun, Moon and Earth.

65. Al-Khalili

Shams al-Dīn Abū ʿAbd Allāh Muḥammad ibn Muḥammad al-Khalīlī

(Arabic: شمس الدين عبد الله محمد بن محمد الخليلي),

(1320–1380),

was a Mamluk-era Syrian astronomer who compiled extensive tables for astronomical use.

Scientific Contributions

Al-Khalili compiled two sets of mathematical tables with 30,000 entries. He carried out experiments to measure tables for the city of Damascus; and augmented them with the measurements by astronomer Ibn Yunus.

Al-Khalili also computed 13,000 entries into his 'Universal Tables' of different mathematical functions which allowed to generate the solutions of standard problems of spherical astronomy, for any given latitude.

In addition to this, Al-Khalili created a 3,000-entry table that gave the direction of the Qibla for all latitudes and longitudes for all the Muslim countries of the 14th century. The values present in al-Khalili's tables have been determined to be accurate up to three or four significant decimal digits. Details of al-Khalili's calculations of the tables did not survive.

Biographical Summary

Al-Khalīlī was born in 1320 AD and he died in 1380. He worked for most of his life as a timekeeper (muwaqqit) at the Umayyad Mosque in Damascus. Little else is known about his life.

66. Al-Damiri

Kamal al-Din Muhammad ibn Musa al-Damiri, also known as Al-Damiri,

(Arabic: كمال الدين محمد بن موسى الدميري),

(1341–1405),

was an Arab Muslim writer from Egypt on zoology, canon law and natural history.

Scientific Contributions

He wrote the first systematic elaboration of the field of zoology. Al-Damiri wrote his book on "Ḥayāt al-ḥayawān al-kubrā", c.1371 (Life of Animals). It analyzes, in alphabetic order, 931 animals mentioned in the Quran, the traditions and the poetical and proverbial Arabic literature: including over 500 prose writers and nearly 200 poets.

The correct spelling of the names of the animals is given with an explanation of their meanings. The use of the animals in medicine; their lawfulness or unlawfulness as food; their position in folklore are discussed; and occasionally sections on political history are also included.

The work exists in three forms. The fullest has been published several times in Egypt; a mediate and a short recension exist in manuscript form.

A part of the book was extracted containing poetic content by al-Suyuti, was translated into Latin by Abraham Ecchelensis (Paris, 1667).

Bochartus in his Hierozoicon (1663) based the work on al-Damiri's work.

There is a translation of the entire book into English by Lieutenant-Colonel Jayakar (Bombay, 1906–1908).

Biographical Summary

Al-Damiri belonged to one of the two towns called Damira near Damietta and spent his life in Egypt. Of the Shafiite school of law, he became professor of tradition in the Rukniyya at Cairo, and also at the mosque al-Azhar. In connection with this work, he wrote a commentary on the Minhāj al-Ṭalibīn of Al-Nawawi.

Al-Damiri was born in Cairo in 1344 AD and he died in Cairo in 1405.

67. Al-Qalqashandi

Shihāb al-Dīn Abū 'l-Abbās Aḥmad ibn 'Alī ibn Aḥmad 'Abd Allāh al-Fazārī al-Shāfiʿī, better known by the epithet al-Qalqashandī,

(Arabic: شهاب الدين أحمد بن علي بن أحمد القلقشندي)

(1355 or 1356 – 1418),

was an Egyptian encyclopedist, polymath and mathematician.

Scientific Contributions

His magnum opus is the voluminous administrative encyclopedia "Ṣubḥ al-Aʿshá fī Ṣināʿat al-Inshāʾ" ('Daybreak for the Night-Blind regarding the Composition of Chancery Documents'). It is a fourteen-volume encyclopedia completed in 1412. It is an administrative manual on geography, political history, natural history, zoology, mineralogy, cosmography, and time measurement. Based on the Masālik al-abṣār fī mamālik al-amṣar of Shihab al-Umari, it has been called one of the final expressions of the genre of administrative literature.

Selections of the book on "Seats of Government" and "Regulations of the Kingdom" from Early Islam to the Mamluks' have been published separately.

The Ṣubḥ al-aʿshā was the first published discussion of the substitution and transposition of ciphers, and the first description of a polyalphabetic cipher, in which each plaintext letter is assigned more than one substitution. The exposition on cryptanalysis included the use of tables of letter frequencies and sets of letters which cannot occur together in one word. It is the first work in history that described cryptology; it described both cryptography and cryptanalysis.

Al-Qalqashandi quoted the text relevant to cryptology from the work of Ibn al-Durayhim (1312–1361) that was once considered lost. Later discoveries in Istanbul"s Sulaimaniyyah Ottoman Archives found the work by Ibn Duraihim, and also works of al-Kindi.

Biographical Summary

A native of the Nile Delta, Al-Qalqashandi became a Scribe of the Scroll (Katib al-Darj), or clerk of the Mamluk chancery in Cairo, Egypt.

Al-Qalqashandi was born in 1355 or 1356 AD and he died in 1418.

68. Ibn al-Majdi

Shihāb al-Dīn ibn al-Majdī

(Arabic: شهاب الدين بن المجدي)

(1359–1447 AD),

was an Egyptian mathematician and astronomer.

Scientific Contributions

Ibn al-Majdī's most important mathematical work was "Book of Substance", a voluminous commentary on the Summary of the Operations of Calculations by Ibn al-Banna'.

Ibn al-Majdī became a highly regarded teacher in many disciplines as well as in the mathematical sciences. Virtually all of his younger contemporaries and immediate successors who were active in astronomy in Cairo were his pupils at one time or another. A prolific and competent writer, Ibn al-Majdī played an important role as a didactic author; his writings were still read and commented upon in Egypt in the late 19th century.

Ibn al-Majdī's numerous astronomical treatises deal with a wide range of topics, including several devoted to the annual ephemerides.

Ibn al-Majdī's two treatises are "Jāmiʿ al-mufīd fī bayān uṣūl al-taqwīm wa-ʾl-mawālīd" and "Ghunyat al-fahīm wa-ʾl-ṭarīq ilā ḥall al-taqwīm".

A third treatise is "al-Durr al-yatīm fī tashīl ṣināʿat al-taqwīm" which contains a set of auxiliary tables for facilitating the calculation of planetary positions. These tables contain numerical entries for the Sun, Moon, and planets.

Ibn al-Majdī wrote treatises on the determination of the lunar crescent visibility, a topic of prime importance to Muslim religious practice since the Islamic calendar is lunar. He dealt with the applied problems of finding the qibla direction.

Ibn al-Majdī's treatise on sexagesimal arithmetic, a topic of fundamental importance for astronomers, was praised by his former pupil Sibṭ al-Māridīnī as being the only satisfactory treatment of the subject known to him.

A noted religious scholar, he nevertheless treated the topic of mathematical astrology in his "al-Jāmiʿ al-mufīd" and even cast a horoscope for a Mamluk amīr.

Biographical Summary

Ibn al-Majdī was born in Cairo in 1359 AD and he died in Cairo in 1447 AD.

Ibn al-Majdī was one of the major Egyptian astronomers during the first half of the 15th century. He occupied the positions of muwaqqit (timekeeper) at al-Azhar Mosque and of "head of the teachers" at the Jānibakiyya madrasa (privately endowed religious college).

Ibn al-Majdī received a traditional religious education in the fields of Quranic studies, the prophetic traditions (ḥadīth), jurisprudence (fiqh), and Arabic grammar and philology. He also became an expert in arithmetic, geometry, the algebra of inheritance, theoretical astronomy (hay'a), and applied astronomy (mīqāt). He learned the latter discipline under Jamāl al-Dīn al-Māridīnī, who had been a student of the celebrated astronomer of Damascus, Ibn al-Shāṭir.

Concluding Remarks

We have presented 68 scientists in the Natural Sciences from the part 1 (AD 610 to 1400) of Islamic Era (AD 610 to 1922). All of them except two are Muslims, an expression of the fact that the era was entirely dominated by the Muslims in all domains of natural sciences.

The study makes explicit that there is a natural affinity between Islam and science because Quran exhorts its readers to a scientific outlook in life by urging them to observe the nature and the universe around them to get to know them. That is the solidly open path to appreciate the truth in Quran and to approach closer to its Speaker.

On the pragmatic aspect, the requirements of religious acts make it necessary to do scientific research in order to meet those requirements. Following are some examples:

- the need to pray five times a day necessitated reliable time keeping giving rise to various sciences of sun dials, researching for sciences of projection to produce equal shadow during each hour. Masajid employed the leading-edge astronomers as time keepers, and further encouraged them towards breakthroughs in astronomical and mathematical sciences;
- the need to determine the direction of Qibla, over the vast Islamic empire, necessitated research in geography, astronomy, spherical geometry, trigonometry, spherical trigonometry, and analytic geometry (combining geometry and algebra);

- the requirement of wudu necessitated water management and aesthetic fountains in the courtyards of magnificent masajid;
- the requirement of Miqat in haj and umrah have necessitated geographical measurements and geometry calculations;
- the spiritual practices of haj and umrah had brought together Muslims into a world-wide congress, including for the scientific research conferencing;
- and the descriptions of Jannah inspired aesthetic sciences of gardening, water fountains, water-powered-clocks, water mills, and other technological marvels.

The natural synergy between Islam and sciences is inherently necessary, complementary, and mutually supportive. This synergy enabled the following absolutely foundational research contributions to the world civilization; these are some illustrative examples given in chronological order.

1. Let us start with Al-Khwarizmi who contributed the complete algorithms for decimal mathematics at a time when the Europeans were happily busy with their abacus. He calculated the circumference of the Earth, and developed an accurate world map. He wrote the treatise Kitāb al-mukhtaṣar fī ḥisāb al-jabr wal-muqābala (Arabic: الكتاب المختصر في حساب الجبر والمقابلةal) which among other topics discusses "Algebra", details of its operations, solution of polynomial and quadratic equations, with applications in trade, surveying, and legal

inheritance. Invention of Algebra allowed mathematics to be applied to itself in a way which nobody had dreamed of, and was an unmistakable revolution in mathematics. The Greeks had not dreamed of such possibilities in mathematics.

2. Al-Jahiz clearly discussed the struggle for existence and the determining factors of natural selection. He also discussed micro evolution. However, the theory of natural selection and evolution is attributed entirely to Darwin.
3. Al-Kindi did the earliest known use of statistical inference and wrote a book on cryptography and cryptanalysis. He invented methods of breaking ciphers using frequency analysis. However, encyclopedia Britannica in its current article on "History of Cryptology" does not even mention Al-Kindi.
4. Ahmad Ibn Yusuf researched "Ratios and Proportions" and worked sequence of calculations. However, the sequence got attributed entirely to Fibonacci.
5. Abu Kamil expanded Al-Khwarizmi's Algebra by introducing irrational numbers into mathematics; solved 8th order equations, equations with irrational numbers, and equations with integer solutions. He also solved sets of non-linear simultaneous equations with three unknown variables. Abu Kamil described a regular pentagon using a 4th order equation, thus laying down the foundation for analytic geometry.

6. Ibn Yunus' contributed methods for determining the time from solar or stellar altitude. Two of these methods together were equivalent to the trigonometric identity
$2 \cos(a)\cos(b) = \cos(a+b) + \cos(a-b)$
which is attributed entirely to Johannes Werner.
7. Ibn Al-Haytham wrote his seven-volume monumental treatise titled كتاب المناظر, "Book of Optics" which was thitherto the only treatment of optics that could genuinely be described as scientific. In comparison, the Greek treatment was pedestrian and erroneous; who, including Euclid and Ptolemy, believed in an "Emission Theory" whereby the eye is the emitter of light. Without the work of Ibn al-Haytham, the Six Books of Optics by Franciscus Aguilonius, works by Christiaan Huygens, and Hering's law of equal innervation, could not have been possible.
8. Ali Ibn Khalaf and Al-Zarqali invented the Universal Astrolabe and its design was copied in the Libros del Saber of Alfonso X of Castille in Spain.
9. Yusuf al-Mu'taman ibn Hud wrote the monumental treatise titled "Kitab al-Istikmal" in which he proved what is now called Ceva's Theorem.
10. Ali Ibn Ridwan astronomically observed the brightest astronomical event in recorded history, the Supernova SN1006. Ibn Sina also observed it in Iran. Some others observed it with

naked eye without collecting astronomical data. SN1006 is not attributed to Ali Ibn Ridwan, though such observations are often attributed to the Astronomer who observed them, like SN 1987A is attributed to Canadian astronomer Ian K. Shelton.

11. Al-Jayyānī researched the Unknown Arcs of a Sphere and wrote a comprehensive treatise on Spherical Trigonometry, presenting formulae for right-handed triangles, the general law of sines, and the solution of a spherical triangle by means of the polar triangle. These and his definition of ratios as numbers and his method of solving a spherical triangle when all sides are unknown are aspects that Regiomontanus later reported in his own works.
12. Al-Zarqālī noted that the path of the center of the primary epicycle is not a circle, as it is for the other planets. Instead, it is approximately oval contributing the existence of non-circular orbits, such as elliptic orbits.
13. Omar Khayyam solved the cubic equations in mathematics by connecting the cubic equation with geometric conic sections, thus further laying a firm foundation for analytic geometry. He provided proofs in algebra, presented methods for extracting n^{th} root of a number, and researched the Binomial theorem. One of Khayyam's predecessors, Al-Karaji, had already discovered the triangular arrangement of the coeffi-

cients of binomial expansions, and Khayyam popularized this triangular array in Iran, so that it is known as Omar Khayyam's triangle. However, later it came to be called as Pascal's triangle. Omar Khayyam proved a connection of Euclid's parallel axiom with the 4th postulate, in an elaborate attempt to prove the parallel axiom itself. This proof clearly showed the possibility of non-Euclidean geometries. However, these geometries now carry the names of Riemann and Gauss, Bolyai, and Lobachevsky.

14. Jabir Ibn Aflah corrected Almagest of Ptolemy in In his book "Iṣlāḥ al-Majisṭi". However, Regiomontanus reported the results in his own book, named "On Triangles".
15. Nur ad-Din al-Bitruji proposed a physical cause of celestial motions in his book of theoretical astronomy and cosmology, (كتاب الهيئة), which was an explicit search for a force like gravity to drive the astronomical bodies in their orbits. Some writer used the material in a treatise on tides, Escorial MS 1636, dated 1192.
16. Ismail Al-Jazari invented the foundational elements in mechanical engineering, and invented clocks and machines that were advanced in automation and robotics. "The Book of Knowledge of Ingenious Mechanical Devices" (Arabic: كتاب في معرفة الحيل الهندسية) is a detailed analysis of 50 machines and how to construct them using fundamentally new concepts in

mechanical engineering, like camshaft, crankshaft, segmental gear, control design, pump with valves, reciprocating piston motion, and various water raising mechanisms. He used hydropower for automation, a technique later used by Leonardo da Vinci. Al-Jazari invented robots for a waitress's work, hand-washing mechanisms, peacock fountains with servants, robotic music band, water clock with drummers with robotic servants, and castle clocks.

17. Al-Urdi wrote Kitāb al-Hayʾa (كتاب الهيئة), a work on theoretical astronomy, in which he proved a mathematical lemma (Urdi Lemma) to generalize Apollonius theorem to allow an equant in an astronomic model to be replaced with an equivalent epicycle that moved around a deferent centered at half the distance to the equant point: a result upon which Copernicus later based his work.

The work of Nicolaus Copernicus has uncanny similarities between his work and the uncited work of Muslim astronomers, including, specifically, Nasir al-Din al-Tusi, Ibn al-Shatir, Muayyad al-Din al-Urdi, and Qutb al-Din al-Shirazi. There are unmistakable similarities in the Tusi couple and Copernicus' geometric method of removing the Equant from mathematical astronomy. *Not only do both of the methods match geometrically, more importantly they both use the same exact lettering system for each vertex; a detail that seems too preternatural.* Moreover, the fact that several other details of his model also mirror the work of other Muslim astronomers, begs the question if Copernicus' work was his own.

Muslim scientists are the giants on whose shoulders present-day natural sciences are built. All the Muslim scientists described in this book are of a giant stature; however, the book includes but only a few from the Muslim science community in Islamic Era. There are innumerable others, and many have been lost to oblivion. There is a wealth of "science" buried in that community and it remains to be extracted from the archives. Researchers will no doubt make further discoveries. Subsequent editions of this book would expand the set of scientists included, as well as additional details about those already covered.

There is at least a three-fold purpose to this book. One is to invite the world science community, in a manner of civilizational dialogue, to celebrate the science giants that Islamic Era has contributed to the growth of science and technology at its foundational level as well as at the level of expanding its frontiers. Another is to remind the Muslims of their love for "science" which every man and woman must acquire; not for worldly dominance, but for a better humanity in a better world. One other objective is to join hands with the rest of humanity by satisfying the upwelling desire of the youth to know the truth about Muslim civilization and the excellence of their pursuit for knowledge: scientific, humanitarian, cultural, civilizational, and spiritual.

It is time for the world to move ahead of the historical biases, religious prejudices, cultural entanglements, and hegemonic aspirations. All people, together, constitute our humanity, and we hold this truth as self-evident that all humans are created with equal value. So, let us all join hands to work together to make science in the service of making every day a wonderful day in every neighborhood of our planet.

www.ingramcontent.com/pod-product-compliance
Lightning Source LLC
Chambersburg PA
CBHW061143120626
46546CB00005B/1900
9781956601046